Philipp Hempel

Constitutive modeling of amorphous thermoplastic polymers with special emphasis on manufacturing processes

Schriftenreihe des Instituts für Mechanik
Karlsruher Institut für Technologie (KIT)

Band 2

Eine Übersicht aller bisher in dieser Schriftenreihe erschienenen Bände
finden Sie am Ende des Buches.

Constitutive modeling of amorphous thermoplastic polymers with special emphasis on manufacturing processes

by
Philipp Hempel

KIT Scientific Publishing

Dissertation, Karlsruher Institut für Technologie (KIT)
Fakultät für Bauingenieur-, Geo- und Umweltwissenschaften, 2016
Tag der mündlichen Prüfung: 13. Mai 2016
Referenten: Prof. Dr.-Ing. habil. Thomas Seelig
Prof. Dr.-Ing. habil. Alexander Lion

Impressum

KIT Scientific Publishing

Karlsruher Institut für Technologie (KIT)
KIT Scientific Publishing
Straße am Forum 2
D-76131 Karlsruhe

KIT Scientific Publishing is a registered trademark of Karlsruhe Institute of Technology. Reprint using the book cover is not allowed.

www.ksp.kit.edu

Print on Demand 2016

ISSN 2363-4936
ISBN 978-3-7315-0550-1
DOI 10.5445/KSP/1000056493

Constitutive modeling of amorphous thermoplastic polymers with special emphasis on manufacturing processes

Zur Erlangung des akademischen Grades eines

DOKTOR-INGENIEURS

von der Fakultät für

Bauingenieur-, Geo- und Umweltwissenschaften
des Karlsruher Institut für Technologie (KIT)

genehmigte

DISSERTATION

von

M.Sc. Philipp Hempel
aus Hannover

Tag der mündlichen Prüfung: 13.05.2016

Hauptreferent: Prof. Dr.-Ing. habil. Thomas Seelig
Korreferent: Prof. Dr.-Ing. habil. Alexander Lion

Karlsruhe 2016

Abstract

This thesis deals with the development of constitutive models for the mechanical behavior of amorphous thermoplastic polymers at large strains. A special emphasis lies on the temperature dependency so that the altered material behavior at high temperatures can be considered. The material exhibits a ductile behavior above the so-called glass transition temperature and is easy to form while it is stiff and brittle at low temperatures. Based on an existing model for glassy polymers below the glass transition temperature, this complex material behavior is modeled by introducing additional constitutive equations accounting for the temperature dependent stiffness, flow strength and hardening at large strains. Two different models to describe the temperature dependent hardening are investigated for this purpose. These models are fitted to own experimental data and the general ability to simulate complex thermomechanical loading conditions realistically is investigated. It is shown that only one of the two models is suited for this purpose which is then used for thermoforming simulations. The results of these simulations are compared to experimental data and it is found that the modeled deformation behavior is quantitatively and qualitatively in good agreement with the experiments.

The good formability of thermoplastic polymers at elevated temperatures is utilized in manufacturing processes in which the material is heated above the glass transition temperature and is easily formed to the desired shape. A rapid cooling may give rise to a "frozen-in" molecular orientation which leads to an altered mechanical behavior in subsequent deformations. This possible initial molecular orientation in a component is also considered in the constitutive model so that the mechanical behavior of a pre-stretched polymer can be described qualitatively. The deformation behavior at heating above the glass transition temperature of two injection molded components is investigated and serves as a computational example. An initial molecular orientation is assumed and mapped to the computational model. The deformation behavior of the real parts is qualitatively well reproduced in the simulations. In a further computational example, the influence of the amount of initial molecular stretch on the behavior in thermoforming simulations is investigated. It is shown that the influence is pronounced and, hence, the molecular pre-orientation should not be neglected in thermoforming simulations.

To implement the developed constitutive models the software tool AceGen is used by which program code is generated and optimized as well as derivatives are calculated automatically. A methodology is developed to automatically derive the algorithmic consistent tangent, for example. This approach is verified by means of simple, exemplary material models with analytical and approximated solutions of the computed derivatives.

Kurzfassung

In der vorliegenden Arbeit werden kontinuumsmechanische Konstitutivmodelle für das mechanische Verhalten von amorphen thermoplastischen Polymeren bei großen Deformationen entwickelt. Dabei wird speziell auf das stark temperaturabhängige Verhalten eingegangen, womit das veränderte Materialverhalten bei hohen Temperaturen berücksichtigt werden kann. Oberhalb der sogenannten Glasübergangstemperatur ist das Materialverhalten duktil und das Polymer ist gut verformbar, wobei es unterhalb dieser Temperatur steif und ggf. spröde ist. Dieses komplexe Materialverhalten wird aufbauend auf einem für Temperaturen unterhalb des Glasübergangs existierenden Materialmodell modelliert, indem zusätzliche konstitutiven Gleichungen eingeführt werden, um die temperaturabhängige elastische Steifigkeit, Fließspannung sowie die Verfestigung bei großen Dehnungen realistisch zu beschreiben. Zwei verschiedene Modelle zur Beschreibung der Verfestigung werden dazu untersucht. Die Modelle werden an eigenen experimentellen Versuchsdaten angepasst und die generelle Eignung zur realistischen Simulation von komplizierten thermomechanischen Belastungen, die z.B. beim Thermoformen auftreten, untersucht. Es wird gezeigt, dass nur eines der Modelle dazu geeignet ist, welches danach in Thermoformsimulationen verwendet wird. Die Ergebnisse dieser Simulationen werden mit experimentellen Daten verglichen, wobei das Deformationsverhalten qualitativ sowie quantitativ gut abgebildet werden kann.

Die gute Verformbarkeit von thermoplastischen Polymeren bei hohen Temperaturen wird in verschiedenen Herstellungsverfahren genutzt, wobei das Material überhalb der Glasübergangstemperatur erwärmt wird und dann leicht in die gewünschte Form gebracht werden kann. Ein schnelles Abkühlen kann jedoch zu einer „eingefrorenen" Molekülorientierung führen, die zu einem veränderten Materialverhalten führt. Diese mögliche initiale Molekülorientierung im Bauteil wird zusätzlich im konstitutiven Modell berücksichtigt, wodurch das Materialverhalten eines vorgestreckten Polymers qualitativ abgebildet werden kann. Das Deformationsverhalten bei Erwärmung überhalb der Glasübergangstemperatur zweier spritzgegossener Bauteile wird untersucht und dient als Berechnungsbeispiel. Dazu wird eine angenommene initiale Molekülorientierung auf das Berechnungsmodel aufgebracht. In den Simulationen kann das Deformationsverhalten der real verformten Teile qualitativ gut wiedergegeben werden. Als weiteres Berechnungsbeispiel dient wiederum das Thermoformen, wobei der Einfluss der Vorstreckung im Polymer auf das Umformverhalten untersucht wird. Es kann dabei gezeigt werden, dass der Einfluss der Vorstreckung sehr ausgeprägt ist und somit eine mögliche Molekülorientierung in Thermoformsimulationen nicht vernachlässigt werden sollte.

Zur Implementierung der entwickelten Materialmodelle wird das Programm AceGen verwendet, mit welchem Programmcode generiert und optimiert werden kann sowie automatisch Ableitungen gebildet werden können. Es wird eine Methode entwickelt, um z.B. die algorithmisch konsistente Tangente automatisch zu bestimmen. Dieses Vorgehen wird anhand von einfacheren, beispielhaften Materialmodellen mit analytischen und approximierten Lösungen der berechneten Ableitungen verifiziert.

Acknowledgements

First and foremost I want to thank my advisor Prof. Thomas Seelig who pushed me into the fascinating field of material modeling of polymers and gave me the possibility of doing research on this topic. This thesis would not have been possible without his outstanding and patient support and his permanent willingness for on-topic (and off-topic) discussions. I really appreciate all his contributions which made my time as a Ph.D. student at the Institute of Mechanics a great experience in which I learnt so much. It has been an honor to be his first Ph.D. student!

I also thank Prof. Alexander Lion for his willingness to be the co-referee and for his very helpful comments on the manuscript. Furthermore, I thank Prof. Marc Kamlah and Prof. Thomas Böhlke who served as additional examiner and Prof. Peter Betsch who served as chairman.

Moreover, I thank all people who somehow contributed to this thesis. In particular, I thank Christoph Schmied and Steffen Mattern for very special AceGen discussions, Martin Helbig for so much on- and off-topic discussions and Willi Wendler for helping me with performing the experiments. I thank Friedemann Streich and Tobias Gnauert for doing some routine piece of work. I very appreciate that Markus Heilig and Matthias Worgull (Institute of Microstructure Technology, KIT) provided an oven for our testing machine, tensile test specimen and their experimental results of thermoforming experiments. In addition, I thank Prof. Arild Clausen and Anne Serine Ognedal for a fruitful and great time at the SIMLab, NTNU in Trondheim.

Lastly, I would like to thank my parents for their support during my studies. And most of all, I really appreciate the support of my loving, encouraging, and patient wife Natalie especially during the final stages of this thesis and, of course, my three boys who keep me awake.

Waldbronn, June 2016
Philipp Hempel

Contents

List of Figures

List of Tables

Nomenclature

$\mathcal{B}_0$, $\mathcal{B}_t$	body associated with reference and current configuration, respectively
$\hat{\mathcal{B}}$, $\tilde{\mathcal{B}}$	body with respect to intermediate configurations
$\boldsymbol{X}$, $\boldsymbol{x}$	position vector of material point with respect to reference configuration $\mathcal{B}_0$ and to current configuration $\mathcal{B}_t$, respectively
$\boldsymbol{F}$	deformation gradient
$\boldsymbol{F}^e$, $\boldsymbol{F}^p$	elastic and inelastic part of deformation gradient
J, J^e, J^p	determinant of $\boldsymbol{F}$, $\boldsymbol{F}^e$, $\boldsymbol{F}^p$
λ_i, λ_i^e, λ_i^p	principal stretches, elastic and inelastic parts, respectivley
$\boldsymbol{u}$	displacement vector
$\boldsymbol{U}$, $\boldsymbol{U}^e$, $\boldsymbol{U}^p$	right stretch tensor, elastic and inelastic parts
$\boldsymbol{V}$, $\boldsymbol{V}^e$, $\boldsymbol{V}^p$	left stretch tensor, elastic and inelastic parts
$\boldsymbol{R}$, $\boldsymbol{R}^e$, $\boldsymbol{R}^p$	rotation tensor, elastic and inelastic parts
$\boldsymbol{F}_N$, $\boldsymbol{F}_N^e$, $\boldsymbol{F}_N^p$	network deformation gradient, elastic and inelastic parts
$\boldsymbol{C}$	right Cauchy-Green tensor
$\hat{\boldsymbol{C}}^e$	elastic right Cauchy-Green tensor with respect to intermediate configuration $\hat{\mathcal{B}}$
$\boldsymbol{b}$	left Cauchy-Green tensor
$\hat{\boldsymbol{b}}^p$	inelastic left Cauchy-Green tensor with respect to intermediate configuration $\hat{\mathcal{B}}$
$\tilde{\boldsymbol{b}}_N^e$	elastic network left Cauchy-Green tensor with respect to intermediate configuration $\tilde{\mathcal{B}}$
$\boldsymbol{E}$	Green-Lagrange strain tensor
$\boldsymbol{e}^H$	logarithmic strain tensor
$\boldsymbol{l}$, $\boldsymbol{l}^e$, $\boldsymbol{l}^p$	velocity gradient, elastic and inelastic parts with respect to current configuration $\mathcal{B}_t$
$\hat{\boldsymbol{l}}^e$, $\hat{\boldsymbol{l}}^p$	elastic and inelastic velocity gradient with respect to intermediate configuration $\hat{\mathcal{B}}$
$\tilde{\boldsymbol{l}}_N^e$, $\tilde{\boldsymbol{l}}_N^p$	elastic and inelastic network velocity gradient with respect to intermediate configuration $\tilde{\mathcal{B}}$

$\boldsymbol{d}, \boldsymbol{d}^e, \boldsymbol{d}^p$	rate of deformation tensor, elastic and inelastic parts with respect to current configuration $\mathcal{B}_t$
$\hat{\boldsymbol{d}}^p$	inelastic rate of deformation tensor with respect to intermediate configuration $\hat{\mathcal{B}}$
$\tilde{\boldsymbol{d}}_N^p$	inelastic network rate of deformation tensor with respect to intermediate configuration $\tilde{\mathcal{B}}$
$\boldsymbol{w}$	spin tensor
$\hat{\boldsymbol{w}}^p$	inelastic spin tensor with respect to intermediate configuration $\hat{\mathcal{B}}$
$\tilde{\boldsymbol{w}}_N^p$	inelastic network spin tensor with respect to intermediate configuration $\tilde{\mathcal{B}}$
$\boldsymbol{F}_s^p$	network state deformation gradient
$\boldsymbol{V}_N^i$	initial network stretch tensor
$\lambda_{N\alpha}^i$	principal initial network stretch
$\boldsymbol{\sigma}$	Cauchy stress tensor
$\boldsymbol{\tau}$	Kirchhoff stress tensor
$\boldsymbol{\Sigma}$	Mandel stress tensor
$\boldsymbol{\Sigma}^*$	driving stress tensor
$\hat{\boldsymbol{\tau}}^b$	backstress tensor with respect to intermediate configuration $\hat{\mathcal{B}}$
$\tilde{\boldsymbol{\tau}}^b$	backstress tensor with respect to intermediate configuration $\tilde{\mathcal{B}}$
$\boldsymbol{S}$	second Piola-Kirchhoff stress tensor
$\hat{\boldsymbol{S}}$	second Piola-Kirchhoff stress tensor with respect to intermediate configuration $\hat{\mathcal{B}}$
$\mathbb{C}$, $\mathbb{c}$	reference and spatial tangent modulus
$\mathbb{C}^{alg}$, $\mathbb{c}^{alg}$	algorithmic consistent tangent with respect to reference and current configuration, respectively
$\boldsymbol{f}$	force vector
p	hydrostatic pressure
$\boldsymbol{R}$	residual vector
$\boldsymbol{X}$	vector including all unknowns
$\boldsymbol{J}$	Jacobian
T	absolute temperature
T_g	glass transition temperature
$(\bullet)_{RT}$	quantity at room temperature
$\dot{\varepsilon}$	strain rate
ε^p	plastic strain
l_0, t_0, w_0, A_0	initial length, thickness, width, cross section

l, t, w, A	current length, thickness, width, cross section
ρ_0, ρ	mass density with respect to reference and current configuration, respectively
V, v	volume with respect to reference and current configuration, respectively
Ψ, Ψ^p, Ψ^p	free energy per unit volume, elastic and inelastic parts
W, W^p, W^p	strain energy function, elastic and inelastic parts
s	entropy
$\mathcal{D}_{int}$	internal dissipation
E, ν	Young's modulus, Poisson's ratio
λ, μ	first Lamé parameter, shear modulus (second Lamé parameter)
$\dot{\gamma}^p$	inelastic shear strain rate
$\dot{\gamma}_0^p$	reference inelastic strain rate
A	material parameter of double-kink model
s_0	shear strength
λ_C	mean chain stretch
λ_L	limit chain stretch
λ_{CL}	ratio of mean chain stretch and limit chain stretch
N	rigid links between entanglements
n	chain density
β	inverse Langevin function
C^R	hardening modulus
C^{R*}	scaled hardening modulus
α	tensile-compression asymmetry parameter
$\Delta T_g, E_{g1}, E_{g2}, E_{r1}, E_{r2}$	material parameters of temperature dependent Young's modulus
A_0, A_T	material parameters of temperature dependent function $\tilde{A}(T)$ used for modification of inelastic shear strain rate
B, D, E_α	material parameters of entanglement dissociation model
$\dot{\gamma}_N^p$	rate of molecular relaxation
λ_N	network stretch
λ_N^L	relaxation limit stretch
$\dot{\gamma}_{N0}^p, A_{NT}, \kappa, \xi, \lambda_{N0}^L, \lambda_{NT}^L, \tilde{\tau}_0^b$	material parameters of molecular relaxation model
δ_{ij}	Kronecker's delta
$\mathbf{1}$	unit tensor

I_A, II_A, III_A	first, second and third invariant of a tensor
$\mathrm{tr}[\boldsymbol{A}]$	trace of a tensor
$\det[\boldsymbol{A}]$	determinant of a tensor
$\mathrm{grad}[\boldsymbol{A}]$	gradient of a tensor
$\mathrm{div}[\boldsymbol{A}]$	divergence of a tensor
$\boldsymbol{A}^T$	transpose of a tensor
$\boldsymbol{A}^{-1}$	inverse of a tensor
$\boldsymbol{A}'$	deviatoric part of a tensor
$\check{\boldsymbol{A}}$	isochoric part of a tensor
$\dot{\boldsymbol{A}}$	time derivative of a tensor
$\Delta(\bullet)$	increment, linearization
$\frac{\partial(\bullet)}{\partial(\bullet)}, \frac{\mathrm{d}(\bullet)}{\mathrm{d}(\bullet)}$	partial, total derivative
$(\bullet)_t$, $(\bullet)_{t+\Delta t}$	quantity at time t (beginning of time step Δt) and $t + \Delta t$ (end of time step)
$(\bullet)_0$	initial quantity at time $t = 0$
$\lVert \bullet \rVert$	Frobenius norm
η	relative error

1. Introduction

1.1. Motivation

Nowadays, products made of polymers are ubiquitous in everyday life. The manifold and outstanding properties display polymers as popular engineering materials. In its neat state, modified with particles or reinforced with fibres various components are realized in virtually every industry. To produce components made of neat amorphous thermoplastic polymers, the raw material or semi-finished parts are typically processed at elevated temperatures due to their easy formability in this temperature region. To improve these manufacturing processes by predicting the relationship between process parameters and the quality of the final part with the aim to design optimal products, modern simulation techniques may be utilized. For this purpose it is necessary to describe the complicated mechanical behavior by realistic constitutive models. The large strain mechanical behavior of amorphous thermoplastic polymers is strongly temperature and strain rate dependent. The behavior drastically alters in the temperature region where the material changes from a glassy to a rubbery state: the so called glass transition temperature T_g. In the glassy state (at low temperatures) thermoplastic polymers exhibit a stiff (and eventually brittle) response while they are highly ductile and easily deformable in the rubbery state. This behavior can be demonstrated by applying a bending load at a test specimen, as shown in Fig. 1.1. At room temperature brittle fracture at small strains without any plastic deformation takes place. At elevated temperatures large plastic deformations could be applied without fracture.

Figure 1.1.: Temperature dependence of mechanical behavior amorphous thermoplastic polymers (PMMA). Brittle fracture at room temperature (top), large plastic deformation at elevated temperatures.

This property of the material is utilized in forming processes like injection molding, extrusion or thermoforming. However, process temperatures above the glass transition temperature and subsequent rapid cooling give rise to incidentally or deliberately frozen-in stretch and orientation of the molecular network in manufactured components. The pre-deformed network in the material affects the mechanical properties, e.g. flow strength, hardening and limiting extensibility. Also the thermomechanical behavior is affected: after heating near or above the glass transition the polymer chains are able to relax to a random state since the flow strength is reduced and consequently the part deforms. This phenomenon can be observed, for example, with the injection molded tensile test specimen made of PMMA shown in Fig. 1.2a. It is clearly observable that after annealing the highest deformation occurred near to the gate where the melt flew into the mold during the injection process (Fig. 1.2b). This indicates that in this region the molecular network is highly stretched. This molecular stretch is then "frozen-in" after rapid cooling.

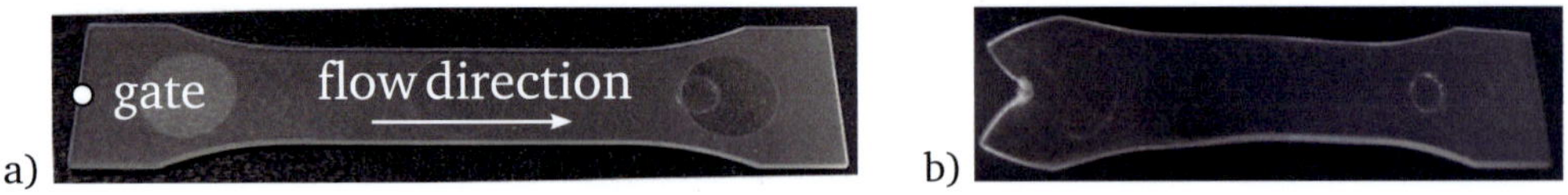

Figure 1.2.: Injection molded tensile test specimen. Initial state and the indicated gate and flow direction (left) and the deformed part after heating (right).

The molecular chain orientations can be made visible with polarized light due to the effect of birefringence. The tensile test specimen is lighted from behind by a polarized light source and photographed through a polarization filter (Fig. 1.3a). In the region at the gate the brightest color is present and with increasing distance from the gate the color changes to a darker one. The bright colors indicate a high level of molecular orientation (frozen-in stretch) which correlates with the deformation seen in Fig. 1.2b. No stretch is visible anymore after heating above T_g because the molecules are relaxed to a random state (Fig. 1.3b) and the initially anisotropic molecular network changed to an isotropic microstructure.

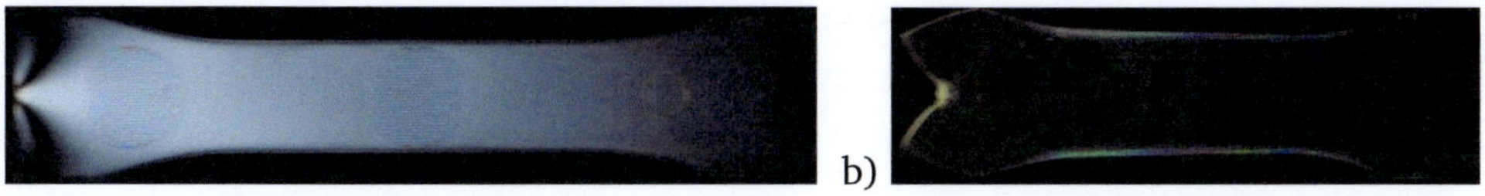
a) b)

Figure 1.3.: Molecular stretch and orientation in the injection molded test specimen made visible due to birefringence under polarized light (left). No orientation is visible after heating above glass transition temperature (right).

Many constitutive models were developed over the last decades accounting for the large strain, rate and temperature dependent behavior of glassy amorphous thermoplastic polymers. The pioneering works of Boyce et al. (1988), Boyce et al. (1989a) are based on the double-kink model (Argon, 1973) to model yielding of the polymer and on the theory of an underlying molecular network to describe the post-yield strain hardening behavior which was originally developed for rubber elasticity (e.g. Wang and Guth (1952), Treloar (1974), Arruda and Boyce (1993), Wu and van der Giessen (1993), Reese (2003), Miehe et al. (2004)). Several groups followed this approach and similar and improved models were developed to describe the deformation behavior of glassy polymers with respect to different loading situations or rates, e.g. Wu and van der Giessen (1994), Hasan and Boyce (1995), Arruda et al. (1995), Tomita and Tanaka (1995), Tervoort et al. (1997), Anand and Gurtin (2003), Govaert and Tervoort (2004), Polanco-Loria et al. (2010), van Breemen et al. (2011), Miehe et al. (2011), Holopainen and Wallin (2013). In the recent years, a couple of new models were developed to capture the change in the mechanical behavior around the glass transition temperature. However, most of these models are only able to reproduce the behavior around or above T_g but not in the glassy state. These models introduce additional equations to describe the change in the plastic flow resistance and micromechanisms thought to be responsible for the temperature dependent hardening at large strains. Two of these micromechanisms at elevated temperatures are "dissociation" of entanglements in the molecular network (Richeton et al., 2007a) and molecular relaxation by "reptation" of chains (e.g. Boyce et al. (2000), Dooling et al. (2002a), Makradi et al. (2005), Palm and Dupaix (2006), Dupaix and Boyce (2007)). Ames et al. (2009) and Srivastava et al. (2010) proposed a rather complex thermomechanical coupled model which is able to reproduce the mechanical behavior from room temperature till above T_g. They also argued with the micromechanism of reptation but introduced a large number of material parameters which makes the model hardly applicable. However, the response of their model is in very good agreement with experimental results (uniaxial compression tests).

The influence of an initial molecular stretch on the material behavior is discussed and modeled in Boyce et al. (1989a) and Arruda et al. (1993) who introduced an initial stretch tensor to incorporate a network pre-stretch into the model. De Focatiis and Buckley (2011) developed a hybrid glass-melt model which combines a molecularly-based melt model with a glassy polymer model. With this model it is possible to consider molecular stretch from the melt phase in the glassy state after cooling below T_g.

An aim of this work is to extend an existing constitutive model (Boyce et al., 1988) to be suitable over a wide temperature range (from room temperature till above the glass transition) while keeping the model as simple as possible with respect to the number of material parameters needed to describe the mechanical behavior sufficiently correct. With the developed model it should be possible to simulate complicated thermomechanical processes such as thermoforming including all process steps. Furthermore, the model should include the possibility of taking a pre-stretch due to preceding processing into account. A mapping of an initial stretch and molecular orientation on the finite element model of a component could save a lot of time in the product development process because the simulation of the whole process chain can be avoided in that case. A model including all these features is not yet available in the literature.

1.2. Outline

The basic continuum mechanics equations needed in this work are given in **Chapter 2**. Starting with the kinematics and stresses, time derivatives and objective rates are discussed as well as the basic principal laws. In addition, the solution of the balance of the linear momentum with the finite element method is explained in Sec. 2.5.

Due to the necessity to understand the mechanical behavior prior to the constitutive modeling the mechanical behavior of amorphous thermoplastic polymers is presented in **Chapter 3**. The microstructure of the polymer and the underlying molecular network as the origin of the mechanical behavior is discussed within a review of experimental results. Special emphasis lies on the strong temperature dependence of the material with the change from a glassy to a rubbery state in the region of the glass transition temperature and on the influence of a pre-stretch on the mechanical behavior. Since many large strain experimental data reported in literature are from compression tests while the main loading condition in forming processes is dominated by tensile strains and in order to obtain experimental data, uniaxial tensile tests on PMMA and PC are conducted with the results presented in Sec. 3.2. The considered temperature range thereby varies from room temperature till above the glass transition temperature. To measure the local strain field a digital image correlation tool is used which is briefly described in Sec. 3.2.2.

Chapter 4 deals with the modeling of the mechanical behavior of amorphous thermoplastic polymers. It starts with a detailed description of the basic model adopted from Boyce et al. (1988). The kinematics and the constitutive equations are established and the thermodynamic consistency of the model is shown. This basic model is extended to be applicable till above the glass transition temperature and to capture the change in the mechanical behavior of the polymer when it converts from the glassy to the rubbery state. Two different models accounting for the above mentioned micromechanisms of "entanglement dissociation" and "reptation" to describe changes in the molecular entanglement network at elevated temperatures are discussed in this context. To compare these models a fictitious thermomechanical loading history is applied and the response of both models in terms of stresses and strains is analyzed. At the end of the chapter a further extension of the model is introduced in which an initial stretch of the material, e.g. due to a manufacturing process, can be accounted for. As computational examples, two injection molded components with an initially applied stretch are investigated with the extended model.

Chapter 5 is related to the validation of the developed material model. For this purpose, a micro-thermoforming process is simulated and the results are compared to experimental results. To investigate the dependence of the initial anisotropy in the polymer film on the forming behavior a thermoforming process of a blister like part is simulated additionally.

Programming of the developed constitutive models is a further part in this work. For this purpose, in **Chapter 6** the implementation of user material models into the finite element program Abaqus using the the automatic differentiation, code generation and optimization tool AceGen is presented. AceGen helps to program the equations very fast and to perform derivatives automatically. Thus, it accelerates the developing and implementation process because the material model developer can focus on essential parts of the work and not on coding or forming complicated derivatives. For the verification of AceGen a hyperelastic and an elastic-plastic material model with a similar structure like the models of Ch. 4 are programmed with AceGen. The automatically computed algorithmic tangent is then compared to the Jacobian derived by two different numerical approximation methods.

Finally, conclusions and an outlook are given in **Chapter 7**.

2. Basics of continuum mechanics

In this chapter the basic equations of non-linear continuum mechanics of solid bodies needed for this work will be given. The formulation of a mechanical (initial) boundary value problem (BVP) with boundary conditions will be defined. The kinematics of a solid body, balance and constitutive principles are formulated. The strong form can be transformed to the weak form of the momentum equation which may be solved with the finite element method (FEM). This procedure will be described briefly in this section, too. More details may be found, e.g., in the textbooks by Belytschko et al. (2000), Parisch (2003), Holzapfel (2007), Gurtin et al. (2010), Wriggers (2008) and Neto et al. (2008).

2.1. Kinematics and strain tensors

The kinematics is the geometrical description of the motion of a body without regarding the cause of the motion. It is necessary to quantify the amount of displacements or deformations of a body. The basic concept of kinematics in the framework of non-linear continuum mechanics is described below and is schematically depicted in Fig. 2.1. In the continuum mechanical theory, a solid body $\mathcal{B}$ is assumed to be a continuous set of material points or particles in the Euclidean space $\mathbb{E}^3$. A body at a certain time t is called a configuration of that body $\mathcal{B}$. At time $t = 0$ the configuration is called the initial configuration and at a fixed reference time the reference configuration. In this work, both configurations coincide with each other and are named reference configuration $\mathcal{B}_0$. The position of a material point in this configuration is defined with the position vector $\boldsymbol{X}$ from the fixed origin of the rectangular coordinate axes. The position vector of the same material point at the time $t > 0$ to the current configuration $\mathcal{B}_t$ is indicated with $\boldsymbol{x}$. The components of the vector $\boldsymbol{X}$ are called the reference or material coordinates and the components of $\boldsymbol{x}$ are the current or spatial coordinates. In the following upper case letters refer mostly to tensor quantities with respect to the reference configuration and to material coordinates. This is called the material or Lagrangian description wherein the motion of a material point is observed. Lower case letters refer to quantities with respect to the current configuration and denote the spatial or Eulerian description where variations in time at a fixed point in space are observed.

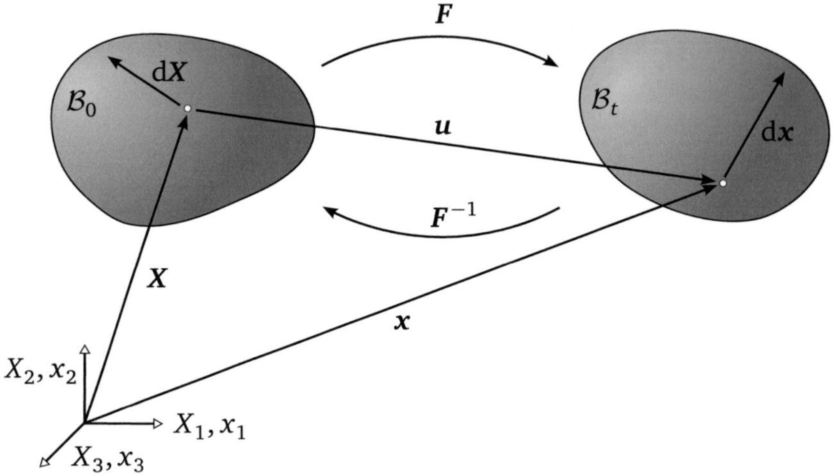

Figure 2.1.: Configurations of a body

The motion of the body can be described as a sequence of configurations in time. With the motion vector field $\boldsymbol{\chi}$ the position of a point in the current configuration may be identified with

$$\boldsymbol{x} = \boldsymbol{\chi}(\boldsymbol{X}, t) \tag{2.1}$$

which is assumed to be invertible

$$\boldsymbol{X} = \boldsymbol{\chi}^{-1}(\boldsymbol{x}, t) \quad . \tag{2.2}$$

The difference between the positions of a material point in the current and the reference configuration is the material displacement field $\boldsymbol{u}(\boldsymbol{X}, t)$ as a function of material coordinates

$$\boldsymbol{u}(\boldsymbol{X}, t) = \boldsymbol{\chi}(\boldsymbol{X}, t) - \boldsymbol{X} = \boldsymbol{x}(\boldsymbol{X}, t) - \boldsymbol{X} \quad . \tag{2.3}$$

A central deformation measure is the deformation gradient $\boldsymbol{F}$. It linearly maps a tangent vector $\mathrm{d}\boldsymbol{X}$ of a material curve (a line of particles) at the point $\boldsymbol{X}$ in $\mathcal{B}_0$ to a tangent vector $\mathrm{d}\boldsymbol{x}$ at the point $\boldsymbol{x}$ in $\mathcal{B}_t$

$$\mathrm{d}\boldsymbol{x} = \boldsymbol{F}\,\mathrm{d}\boldsymbol{X} \quad \text{with} \quad \boldsymbol{F} = \frac{\partial \boldsymbol{\chi}(\boldsymbol{X}, t)}{\partial \boldsymbol{X}} = \frac{\partial \boldsymbol{x}(\boldsymbol{X}, t)}{\partial \boldsymbol{X}} \quad . \tag{2.4}$$

The tangent vector is also called line element: in the reference configuration the material line element $\mathrm{d}\boldsymbol{X}$, respectively, in the current configuration the spatial line element $\mathrm{d}\boldsymbol{x}$. The inversion of the mapping is given by $\boldsymbol{F}^{-1}$ while the determinant $J = \det[\boldsymbol{F}]$ of the deformation gradient must be non zero at all time. Physically, the determinant must be positive to prevent self intersection of the body. A further transformation of a differential quantity from the reference to the current configuration is given by Nanson's formula

$$\mathrm{d}a\, \boldsymbol{n} = \mathrm{d}\boldsymbol{a} = J\ \boldsymbol{F}^{-T}\ \mathrm{d}\boldsymbol{A} = J\ \boldsymbol{F}^{-T}\ \mathrm{d}A\, \boldsymbol{N} \tag{2.5}$$

with the surface elements $\mathrm{d}A$ and $\mathrm{d}a$ and the unit normal vectors $\boldsymbol{N}$and $\boldsymbol{n}$ in the respective configurations. A volume element is transformed by the determinant of the deformation gradient which is the volume ratio

$$\mathrm{d}v = J\ \mathrm{d}V \quad . \tag{2.6}$$

For material modeling it is sometimes useful to use an isochoric deformation gradient $\check{\boldsymbol{F}}$ so that $\det[\check{\boldsymbol{F}}] = 1$ holds. With this requirement one may calculate the isochoric deformation gradient from the total deformation gradient (2.4) as

$$\check{\boldsymbol{F}} = c\ \boldsymbol{F} \ \longrightarrow\ \det[\check{\boldsymbol{F}}] = \det[c\ \boldsymbol{F}] = c^3 \det[\boldsymbol{F}] = 1 \quad . \tag{2.7}$$

So, it is $c = J^{-1/3}$ and the isochoric deformation gradient reads

$$\check{\boldsymbol{F}} = J^{-1/3}\ \boldsymbol{F} \quad . \tag{2.8}$$

The above equation represents a multiplicative split of the total deformation into a volume changing and a volume preserving part. Furthermore, the deformation gradient may be polar decomposed into a pure stretch and a pure rotation

$$\boldsymbol{F} = \boldsymbol{R}\boldsymbol{U} = \boldsymbol{V}\boldsymbol{R} \tag{2.9}$$

with the symmetric stretch tensors:

- the right stretch tensor $\boldsymbol{U}$ defined in the reference configuration
- the left stretch tensor $\boldsymbol{V}$ defined in the current configuration

and the orthogonal rotation tensor $\boldsymbol{R}$. The spectral decomposition of the stretch tensors

$$\boldsymbol{U} = \sum_{a=1}^{3} \lambda_a \, \boldsymbol{N}_a \otimes \boldsymbol{N}_a \tag{2.10}$$

$$\boldsymbol{V} = \sum_{a=1}^{3} \lambda_a \, \boldsymbol{n}_a \otimes \boldsymbol{n}_a \tag{2.11}$$

gives the principal stretches λ_a which are the eigenvalues and are equal for both tensors. Only the eigenvectors differ in the rotation tensor

$$\boldsymbol{n}_a = \boldsymbol{R}\boldsymbol{N}_a \quad . \tag{2.12}$$

In the principal directions of the stretch tensors, the principal stretches are given by the current length l of a line element with respect to the initial length l_0

$$\lambda = \frac{l}{l_0} \quad . \tag{2.13}$$

It is necessary to exclude the rotation to measure the strain because a rigid body motion induces no stress. For this purpose, the rotation free strain measures, i.e. the right Cauchy-Green tensor $\boldsymbol{C}$ and the left Cauchy-Green tensor $\boldsymbol{b}$ are defined as

$$\boldsymbol{C} = \boldsymbol{F}^T\boldsymbol{F} = \boldsymbol{U}\boldsymbol{R}^T\boldsymbol{R}\boldsymbol{U} = \boldsymbol{U}^2 = \sum_{a=1}^{3} \lambda_a^2 \, \boldsymbol{N}_a \otimes \boldsymbol{N}_a \tag{2.14}$$

$$\boldsymbol{b} = \boldsymbol{F}\boldsymbol{F}^T = \boldsymbol{V}\boldsymbol{R}\boldsymbol{R}^T\boldsymbol{V} = \boldsymbol{V}^2 = \sum_{a=1}^{3} \lambda_a^2 \, \boldsymbol{n}_a \otimes \boldsymbol{n}_a \quad . \tag{2.15}$$

To get zero strain when no deformation of the body occurred, two strain tensors are defined: the Green-Lagrange tensor with respect to the reference configuration

$$\boldsymbol{E} = \frac{1}{2}(\boldsymbol{C} - \boldsymbol{1}) \tag{2.16}$$

and the spatial Euler-Almansi strain tensor

$$\boldsymbol{e} = \frac{1}{2}(\boldsymbol{1} - \boldsymbol{b}^{-1}) \quad . \tag{2.17}$$

With the stretch tensors (2.14) and (2.15) generalized strain tensors may be defined in the reference and the current configuration

$$\boldsymbol{E}^{\alpha} = \frac{1}{\alpha}(\boldsymbol{U}^{\alpha} - \mathbf{1}) \quad , \quad \alpha \in \mathbb{R} \tag{2.18}$$

$$\boldsymbol{e}^{\alpha} = \frac{1}{\alpha}(\boldsymbol{V}^{\alpha} - \mathbf{1}) \quad , \quad \alpha \in \mathbb{R} \quad . \tag{2.19}$$

For example, the tensors (2.16) and (2.17) results from (2.18) for $\alpha = 2$ and from (2.19) for $\alpha = -2$, respectively. For $\alpha = 0$ the logarithmic strain tensors, named as Hencky strain tensors, are defined as

$$\boldsymbol{E}^{H} = \ln \boldsymbol{U} = \sum_{a=1}^{3} \ln \lambda_a \, \boldsymbol{N}_a \otimes \boldsymbol{N}_a \tag{2.20}$$

$$\boldsymbol{e}^{H} = \ln \boldsymbol{V} = \sum_{a=1}^{3} \ln \lambda_a \, \boldsymbol{n}_a \otimes \boldsymbol{n}_a \quad . \tag{2.21}$$

2.2. Stress measures

In this section, stress tensors are introduced. Various stress tensors may be defined but only the most important ones are presented in the following.

The Cauchy or true stress vector $\boldsymbol{t}$ is defined by the resultant internal force $\mathrm{d}\boldsymbol{f}$ acting on a surface element $\mathrm{d}a$ which lies on a cut through the body in the current configuration

$$\mathrm{d}\boldsymbol{f} = \boldsymbol{t} \; \mathrm{d}a \quad . \tag{2.22}$$

The stress vector depends on the position $\boldsymbol{x}$, the time t and the normal vector $\boldsymbol{n}$ pointing outwards from the surface element $\mathrm{d}a$ and consequently on the orientation of the cut through the body (Cauchy's postulate)

$$\boldsymbol{t} = \boldsymbol{t}(\boldsymbol{x}, t, \boldsymbol{n}) \quad . \tag{2.23}$$

Cauchy's theorem states that the stress vector $\boldsymbol{t}$ may be written as a linear function of the Cauchy stress tensor $\boldsymbol{\sigma}$ and the normal vector $\boldsymbol{n}$

$$\boldsymbol{t}(\boldsymbol{x}, t, \boldsymbol{n}) = \boldsymbol{\sigma}(\boldsymbol{x}, t) \, \boldsymbol{n} \quad . \tag{2.24}$$

Note, the Cauchy stress is independent of the normal vector $\boldsymbol{n}$ and is only a function of $\boldsymbol{x}$ and t. It is also called the true stress because the actual force is referred to the current surface. If the force is related to a surface in the reference configuration the first Piola-Kirchhoff or nominal stress vector $\boldsymbol{T}$ is defined as

$$\mathrm{d}\boldsymbol{f} = \boldsymbol{T}(\boldsymbol{X}, t, \boldsymbol{N})\,\mathrm{d}A \tag{2.25}$$

with the normal vector $\boldsymbol{N}$ pointing outwards from the surface element dA in the reference configuration. Analogous to (2.24) the two-point first Piola-Kirchhoff stress tensor $\boldsymbol{P}$

$$\boldsymbol{T} = \boldsymbol{P}\boldsymbol{N} \tag{2.26}$$

may be defined and is in general not symmetric while the Cauchy stress tensor is symmetric (cf. (2.73)). Using (2.5) one gets the relation between $\boldsymbol{P}$ and $\boldsymbol{\sigma}$

$$\boldsymbol{P} = J\,\boldsymbol{\sigma}\boldsymbol{F}^{-T} \quad . \tag{2.27}$$

To obtain a stress tensor defined purely in the reference configuration the symmetric second Piola-Kirchhoff stress tensor $\boldsymbol{S}$ is defined but has no physical meaning. The definition follows from a pull-back of the Cauchy stress tensor: the transformation from the current to the reference configuration

$$\boldsymbol{S} = J\,\boldsymbol{F}^{-1}\,\boldsymbol{\sigma}\,\boldsymbol{F}^{-T} \tag{2.28}$$

by using the inverse of the deformation gradient. According to the push-forward -the transformation from the reference to the current configuration- of the second Piola-Kirchhoff stress tensor the Kirchhoff stress tensor $\boldsymbol{\tau}$ is obtained

$$\boldsymbol{\tau} = \boldsymbol{F}\,\boldsymbol{S}\,\boldsymbol{F}^{T} = J\,\boldsymbol{\sigma} \quad . \tag{2.29}$$

A stress tensor may be additively decomposed into a hydrostatic and deviatoric part. Exemplary, the Cauchy stress tensor reads

$$\boldsymbol{\sigma} = \boldsymbol{\sigma}' - p\boldsymbol{1} \quad \Leftrightarrow \quad \boldsymbol{\sigma}' = \boldsymbol{\sigma} + p\boldsymbol{1} \tag{2.30}$$

wherein $p = -\dfrac{1}{3}\,\mathrm{tr}[\boldsymbol{\sigma}]$ is the hydrostatic pressure and $\boldsymbol{\sigma}'$ is the deviatoric stress tensor.

2.3. Time derivatives and objective rates

Time derivatives of tensor fields are needed, e.g. to describe time- and/or history dependent material behavior. Because of the time independence of the material coordinates $\boldsymbol{X}$ the material time derivative of material tensor fields like the material displacement field (2.3) is a derivative with respect to the time only. It yields the material velocity field $\boldsymbol{v}(\boldsymbol{X},t)$

$$\frac{\mathrm{d}}{\mathrm{d}t}\boldsymbol{u}(\boldsymbol{X},t) = \frac{\partial \boldsymbol{u}}{\partial t} = \dot{\boldsymbol{u}} = \boldsymbol{v} \tag{2.31}$$

and analogously the material acceleration field $\boldsymbol{a}(\boldsymbol{X},t)$

$$\frac{\mathrm{d}}{\mathrm{d}t}\boldsymbol{v}(\boldsymbol{X},t) = \frac{\partial^2 \boldsymbol{u}}{\partial t^2} = \ddot{\boldsymbol{u}} = \boldsymbol{a} \quad . \tag{2.32}$$

The material time derivative of a spatial tensor field, for example the spatial velocity field $\boldsymbol{v}(\boldsymbol{x},t)$ ($\boldsymbol{x}$ is time dependent) is the spatial acceleration field $\boldsymbol{a}(\boldsymbol{x},t)$ and is given by

$$\frac{\mathrm{d}}{\mathrm{d}t}\boldsymbol{v}(\boldsymbol{x},t) = \frac{\partial \boldsymbol{v}}{\partial t} + \frac{\partial \boldsymbol{v}}{\partial \boldsymbol{x}}\frac{\partial \boldsymbol{x}}{\partial t} = \frac{\partial \boldsymbol{v}}{\partial t} + \mathrm{grad}[\boldsymbol{v}]\;\boldsymbol{v} = \boldsymbol{a}(\boldsymbol{x},t) \tag{2.33}$$

with the relation $\mathrm{grad}[\boldsymbol{v}] = \dfrac{\partial \boldsymbol{v}}{\partial \boldsymbol{x}}$ and the spatial velocity field itself

$$\boldsymbol{v}(\boldsymbol{x},t) = \frac{\partial \boldsymbol{x}}{\partial t} = \dot{\boldsymbol{x}} \quad . \tag{2.34}$$

The material time derivative of the deformation gradient results from (2.4) and (2.31)

$$\frac{\mathrm{d}}{\mathrm{d}t}\boldsymbol{F}(\boldsymbol{X},t) = \dot{\boldsymbol{F}} = \frac{\partial}{\partial t}\left(\frac{\partial \boldsymbol{x}}{\partial \boldsymbol{X}}\right) = \frac{\partial}{\partial \boldsymbol{X}}\left(\frac{\partial \boldsymbol{x}}{\partial t}\right) = \mathrm{Grad}[\boldsymbol{v}(\boldsymbol{X},t)] \tag{2.35}$$

with the material velocity gradient

$$\mathrm{Grad}[\boldsymbol{v}(\boldsymbol{X},t)] = \frac{\partial \boldsymbol{v}(\boldsymbol{X},t)}{\partial \boldsymbol{X}} \tag{2.36}$$

in which the relation $\boldsymbol{v}(\boldsymbol{X},t) = \boldsymbol{v}(\boldsymbol{x},t)$ is used. The spatial velocity gradient is denoted as $\boldsymbol{l}(\boldsymbol{x},t)$ and is defined by (cf. (2.33))

$$\boldsymbol{l}(\boldsymbol{x},t) = \frac{\partial \boldsymbol{v}(\boldsymbol{x},t)}{\partial \boldsymbol{x}} = \mathrm{grad}[\boldsymbol{v}] \tag{2.37}$$

and may be additively split into a symmetric and a skew-symmetric part

$$\boldsymbol{l} = \boldsymbol{d} + \boldsymbol{w} \tag{2.38}$$

$$\boldsymbol{d} = \frac{1}{2}(\boldsymbol{l} + \boldsymbol{l}^T) \tag{2.39}$$

$$\boldsymbol{w} = \frac{1}{2}(\boldsymbol{l} - \boldsymbol{l}^T) \quad , \tag{2.40}$$

which provides the rate of deformation tensor $\boldsymbol{d}$ and the spin tensor $\boldsymbol{w}$. The connection between the material and spatial velocity gradient may be computed via the chain rule

$$\boldsymbol{l} = \frac{\partial \dot{\boldsymbol{x}}}{\partial \boldsymbol{X}} \frac{\partial \boldsymbol{X}}{\partial \boldsymbol{x}} = \frac{\partial}{\partial t}\left(\frac{\partial \boldsymbol{x}}{\partial \boldsymbol{X}}\right) \boldsymbol{F}^{-1} = \dot{\boldsymbol{F}} \boldsymbol{F}^{-1} \tag{2.41}$$

and it follows the important differential equation

$$\dot{\boldsymbol{F}} = \boldsymbol{l} \boldsymbol{F} \quad . \tag{2.42}$$

Another important time derivative is that of the Green-Lagrange strain tensor (2.16)

$$\dot{\boldsymbol{E}} = \frac{1}{2}\dot{\boldsymbol{C}} = \frac{1}{2}(\dot{\boldsymbol{F}}^T \boldsymbol{F} + \boldsymbol{F}^T \dot{\boldsymbol{F}}) = \frac{1}{2}(\boldsymbol{F}^T \boldsymbol{l}^T \boldsymbol{F} + \boldsymbol{F}^T \boldsymbol{l} \boldsymbol{F}) = \boldsymbol{F}^T \boldsymbol{d} \boldsymbol{F} \quad . \tag{2.43}$$

which gives is the pull-back of the rate of deformation tensor $\boldsymbol{d}$.

The time derivative of stress tensors must be performed carefully according to their definition. While the material time derivative of a material stress tensor is defined by (2.31) and is for example for the second Piola-Kirchhoff stress tensor (2.28)

$$\dot{\boldsymbol{S}} = \frac{\partial \boldsymbol{S}(\boldsymbol{X}, t)}{\partial t} \tag{2.44}$$

the material time derivative of spatial stress tensors is not objective. The objectivity, i.e. the invariance under the change of an observer, must be provided for example for constitutive modeling reasons. For this purpose, objective material time derivatives may be constructed making use of the so called Lie derivative. In this derivative a spatial stress field is pulled-back to the reference configuration where the time derivative is performed and the result is pushed-forward to the current configuration. Doing so for the stress tensors in the current configuration one obtains the Lie time derivative of the Cauchy stress tensor as

$$\begin{aligned} \mathfrak{L}(\boldsymbol{\sigma}) &= \boldsymbol{F}\frac{\mathrm{d}(\boldsymbol{F}^{-1}\boldsymbol{\sigma}\boldsymbol{F}^{-T})}{\mathrm{d}t}\boldsymbol{F}^T = \boldsymbol{F}\frac{\mathrm{d}(J^{-1}\boldsymbol{S})}{\mathrm{d}t}\boldsymbol{F}^T \\ &= \boldsymbol{F}(-\frac{\dot{J}}{J^2}\boldsymbol{S} + J^{-1}\dot{\boldsymbol{S}})\boldsymbol{F}^T = -\frac{\dot{J}}{J^2}\boldsymbol{\tau} + J^{-1}\boldsymbol{F}\dot{\boldsymbol{S}}\boldsymbol{F}^T \end{aligned} \tag{2.45}$$

and of the Kirchhoff stress tensor

$$\mathfrak{L}(\boldsymbol{\tau}) = \boldsymbol{F}\frac{\mathrm{d}(\boldsymbol{F}^{-1}\boldsymbol{\tau}\boldsymbol{F}^{-T})}{\mathrm{d}t}\boldsymbol{F}^T = \boldsymbol{F}\frac{\mathrm{d}\boldsymbol{S}}{\mathrm{d}t}\boldsymbol{F}^T = \boldsymbol{F}\dot{\boldsymbol{S}}\boldsymbol{F}^T \quad . \tag{2.46}$$

The definition of the objective Oldroyd stress rate of the Cauchy stress

$$\overset{\triangle}{\boldsymbol{\sigma}} = \dot{\boldsymbol{\sigma}} - \boldsymbol{l}\boldsymbol{\sigma} - \boldsymbol{\sigma}\boldsymbol{l}^T \tag{2.47}$$

follows directly from the first term of (2.45) by applying the product rule

$$\mathfrak{L}(\boldsymbol{\sigma}) = \boldsymbol{F}\frac{\mathrm{d}(\boldsymbol{F}^{-1}\boldsymbol{\sigma}\boldsymbol{F}^{-T})}{\mathrm{d}t}\boldsymbol{F}^T = \boldsymbol{F}\left(\dot{\overline{\boldsymbol{F}^{-1}}}\boldsymbol{\sigma}\boldsymbol{F}^{-T} + \boldsymbol{F}^{-1}\dot{\boldsymbol{\sigma}}\boldsymbol{F}^{-T} + \boldsymbol{F}^{-1}\boldsymbol{\sigma}\dot{\overline{\boldsymbol{F}^{-T}}}\right)\boldsymbol{F}^T \tag{2.48}$$

and using the relations (2.41), $\dot{\overline{\boldsymbol{F}^{-1}}} = -\boldsymbol{F}^{-1}\boldsymbol{l}$ and $\dot{\overline{\boldsymbol{F}^{-T}}} = -\boldsymbol{l}^T\boldsymbol{F}^{-T}$. In the same manner, the Oldroyd stress rate of the Kirchhoff stress

$$\overset{\triangle}{\boldsymbol{\tau}} = \dot{\boldsymbol{\tau}} - \boldsymbol{l}\boldsymbol{\tau} - \boldsymbol{\tau}\boldsymbol{l}^T \tag{2.49}$$

may be found. According to (2.46) the push-forward of the time derivative of the second Piola-Kirchhoff stress tensor is equal to the Oldroyd stress rate of the Kirchhoff stress tensor

$$\overset{\triangle}{\boldsymbol{\tau}} = \boldsymbol{F}\dot{\boldsymbol{S}}\boldsymbol{F}^T \quad . \tag{2.50}$$

To express the above equation in spatial quantities only, the incremental constitutive law

$$\dot{\boldsymbol{S}} = \frac{1}{2}\mathbb{C} : \dot{\boldsymbol{C}} \tag{2.51}$$

with the reference tangent modulus $\mathbb{C} = 2\frac{\partial \boldsymbol{S}}{\partial \boldsymbol{C}}$ is used in combination with (2.43). It follows the relation for the Oldroyd stress

$$\overset{\triangle}{\boldsymbol{\tau}} = \mathbb{c} : \boldsymbol{d} \tag{2.52}$$

where the spatial tangent modulus $\mathbb{c}$ is obtained by the push-forward of $\mathbb{C}$

$$\mathbb{c}_{ijkl} = F_{iI}F_{jJ}F_{kK}F_{lL}\mathbb{C}_{IJKL} \quad . \tag{2.53}$$

Another important objective stress rate is the Jaumann-Zaremba stress rate of the Cauchy stress $\overset{\circ}{\boldsymbol{\sigma}}$. It is defined by setting the rate of deformation to zero ($\boldsymbol{d} = \boldsymbol{0} \Rightarrow \boldsymbol{l} = \boldsymbol{w}$) in the Lie time derivative (2.48) from which it follows

$$\mathfrak{L}(\boldsymbol{\sigma})_{d=0} = \overset{\circ}{\boldsymbol{\sigma}} = \dot{\boldsymbol{\sigma}} - \boldsymbol{w}\boldsymbol{\sigma} + \boldsymbol{\sigma}\boldsymbol{w} \quad . \tag{2.54}$$

2.4. Basic balance principles

In this section the basic balance principles of classical continuum mechanics and thermodynamics are presented which in addition to the kinematics are necessary to mathematically formulate an initial boundary value problem of the continuum. These balance equations are independent of the material and are axioms. The mechanical balance equations are the balance of mass, of linear momentum and angular momentum. The thermodynamical balance equations are the energy balance and the entropy balance. The energy balance is also known as first law of thermodynamics. The second law of thermodynamics states that the entropy production is non-negative. The equations may be expressed in integral form for the whole body or in local form for a differential volume element.

2.4.1. Balance of mass

The scalar valued mass m of a continuum body $\mathcal{B}$, a closed system, is given by the integral of the mass density over the body in the respective configuration of the body

$$m = \int_{\mathcal{B}_0} \rho_0(\boldsymbol{X}) \, \mathrm{d}V = \int_{\mathcal{B}_t} \rho(\boldsymbol{x}, t) \, \mathrm{d}v \tag{2.55}$$

and must be constant at all times such that

$$\dot{m} = \frac{\mathrm{d}}{\mathrm{d}t} \int_{\mathcal{B}_t} \rho(\boldsymbol{x}, t) \, \mathrm{d}v = 0 \tag{2.56}$$

must hold. From the integral forms (2.55) and (2.56) the local forms for the differential volume element may be found because of the validness for all volumes. Using the volume transformation (2.6) to change the variable of the integral on the right hand side in (2.55) leads to

$$\int_{\mathcal{B}_0} (\rho_0(\boldsymbol{X}) - \rho(\boldsymbol{x},t)\, J(\boldsymbol{X},t))\, \mathrm{d}V = 0 \tag{2.57}$$

which gives the local form of the continuity equation

$$\rho_0 = \rho J \quad . \tag{2.58}$$

Analogously, a local form of (2.56) may be obtained by applying the time derivative in the reference configuration and transform the result back to the current configuration

$$\dot{m} = \frac{\mathrm{d}}{\mathrm{d}t} \int_{\mathcal{B}_0} \rho J \; \mathrm{d}V = \int_{\mathcal{B}_0} (\dot{\rho} J + \rho \dot{J})\; \mathrm{d}V = \int_{\mathcal{B}_t} (\dot{\rho} + \rho \; \mathrm{div}[\boldsymbol{v}])\; \mathrm{d}v = 0 \tag{2.59}$$

with the material time derivative of J

$$\dot{J} = J \; \mathrm{div}[\boldsymbol{v}] \quad . \tag{2.60}$$

Thus, the local rate form of continuity equation is

$$\dot{\rho} + \rho \; \mathrm{div}[\boldsymbol{v}] = 0 \quad . \tag{2.61}$$

If a material is incompressible then the density is constant and it follows from (2.61) that the motion must be isochoric

$$\mathrm{div}[\boldsymbol{v}] = \mathrm{tr}[\boldsymbol{d}] = 0 \quad . \tag{2.62}$$

2.4.2. Balance of linear momentum

The linear momentum of a material body is defined by the integral of the product of the mass density and the velocity over the body in the reference or in the current configuration, respectively

$$\boldsymbol{I}(t) = \int_{\mathcal{B}_0} \rho_0(\boldsymbol{X})\; \boldsymbol{v}(\boldsymbol{X},t)\; \mathrm{d}V = \int_{\mathcal{B}_t} \rho(\boldsymbol{x})\; \boldsymbol{v}(\boldsymbol{x},t)\; \mathrm{d}v \quad . \tag{2.63}$$

Based on Newton's second law of motion, the time derivative of (2.63) is equal to the sum of all forces $\boldsymbol{f}(t)$ acting on the body. In spatial coordinates, these forces are separated in forces acting on the surface of the body $\partial \mathcal{B}$, summarized by the Cauchy stress vector $\boldsymbol{t}$

and forces acting on the volume of the body, summarized by the body force vector field $\boldsymbol{b}$ per unit current volume (e.g. gravity load per unit volume)

$$\boldsymbol{f}(t) = \int_{\mathcal{B}_t} \boldsymbol{b}\, \mathrm{d}v + \int_{\partial\mathcal{B}_t} \boldsymbol{t}\, \mathrm{d}a \quad . \tag{2.64}$$

The linear momentum balance equation in the current configuration reads as

$$\dot{\boldsymbol{I}} = \frac{\mathrm{d}}{\mathrm{d}t} \int_{\mathcal{B}_t} \rho \boldsymbol{v}\, \mathrm{d}v = \boldsymbol{f} \quad . \tag{2.65}$$

To perform the time derivative, the integral may be transformed to the reference configuration and by using (2.60) and (2.61) it yields

$$\frac{\mathrm{d}}{\mathrm{d}t} \int_{\mathcal{B}_t} \rho \boldsymbol{v}\, \mathrm{d}v = \frac{\mathrm{d}}{\mathrm{d}t} \int_{\mathcal{B}_0} \rho \boldsymbol{v} J\, \mathrm{d}V = \int_{\mathcal{B}_0} (\dot{\rho} \boldsymbol{v} J + \rho \dot{\boldsymbol{v}} J + \rho \boldsymbol{v} \dot{J})\, \mathrm{d}V = \int_{\mathcal{B}_t} \rho \dot{\boldsymbol{v}}\, \mathrm{d}v \quad . \tag{2.66}$$

Hence, the balance equation may be rewritten by additionally using the divergence theorem as

$$\int_{\mathcal{B}_t} \rho \dot{\boldsymbol{v}}\, \mathrm{d}v = \int_{\mathcal{B}_t} \boldsymbol{b}\, \mathrm{d}v + \int_{\mathcal{B}_t} \mathrm{div}[\boldsymbol{\sigma}]\, \mathrm{d}v \tag{2.67}$$

which is the equation of motion and must hold for all volumes. This leads to the local form

$$\mathrm{div}[\boldsymbol{\sigma}] + \boldsymbol{b} = \rho \dot{\boldsymbol{v}} \quad , \quad \frac{\partial \sigma_{ij}}{\partial x_j} + b_i = \rho \dot{v}_i \tag{2.68}$$

with the inertia term on the right hand side. If this term vanishes because of negligible accelerations then the special case of statics is obtained.

2.4.3. Balance of angular momentum

Besides the mass and the linear momentum balance, also the balance of the angular momentum must be satisfied for the continuum body for all times. The angular momentum is defined relative to a fixed point in space $\boldsymbol{x}_0$ as

$$\boldsymbol{L} = \int_{\mathcal{B}_0} \boldsymbol{r} \times \rho_0(\boldsymbol{X}, t)\, \boldsymbol{v}(\boldsymbol{X}, t)\, \mathrm{d}V = \int_{\mathcal{B}_t} \boldsymbol{r} \times \rho(\boldsymbol{x}, t)\, \boldsymbol{v}(\boldsymbol{x}, t)\, \mathrm{d}v \quad . \tag{2.69}$$

with the position vector $\boldsymbol{r}(\boldsymbol{x}) = \boldsymbol{x} - \boldsymbol{x}_0$. Analogously to the balance of the linear momentum, the change in time of the angular momentum is equal to the sum of all moments acting on the body. These moments result from the forces acting on the body with respect to $\boldsymbol{x}_0$. The angular moment balance in the current configuration reads as

$$\dot{\boldsymbol{L}} = \frac{\mathrm{d}}{\mathrm{d}t} \int\limits_{\mathcal{B}_t} \boldsymbol{r} \times \rho \boldsymbol{v} \; \mathrm{d}v = \int\limits_{\mathcal{B}_t} \boldsymbol{r} \times \rho \dot{\boldsymbol{v}} \; \mathrm{d}v = \int\limits_{\mathcal{B}_t} \boldsymbol{r} \times \boldsymbol{b} \; \mathrm{d}v + \int\limits_{\partial \mathcal{B}_t} \boldsymbol{r} \times \boldsymbol{t} \; \mathrm{d}a \tag{2.70}$$

by evaluating the time derivative as in (2.66). Using the divergence theorem and the permutation tensor $\boldsymbol{\mathcal{E}}$, the second term on the right hand side of (2.70) may be rewritten as an integral over the volume

$$\int\limits_{\partial \mathcal{B}_t} \boldsymbol{r} \times \boldsymbol{t} \; \mathrm{d}a = \int\limits_{\mathcal{B}_t} (\boldsymbol{r} \times \mathrm{div}[\boldsymbol{\sigma}] + \boldsymbol{\mathcal{E}} : \boldsymbol{\sigma}^T) \; \mathrm{d}v \quad . \tag{2.71}$$

Inserting this result in (2.70) and rearranging, it yields

$$\int\limits_{\mathcal{B}_t} \boldsymbol{r} \times (\rho \dot{\boldsymbol{v}} - \boldsymbol{b} - \mathrm{div}[\boldsymbol{\sigma}]) \; \mathrm{d}v = \int\limits_{\mathcal{B}_t} \boldsymbol{\mathcal{E}} : \boldsymbol{\sigma}^T \; \mathrm{d}v \quad . \tag{2.72}$$

With (2.68) it follows the double contraction of the permutation tensor and the transpose of the Cauchy stress tensor must be equal to zero

$$\boldsymbol{\mathcal{E}} : \boldsymbol{\sigma}^T = \boldsymbol{0} \tag{2.73}$$

which only can be true if the Cauchy stress tensor is symmetric.

2.4.4. First law of thermodynamics

The first law of thermodynamics postulates the existence of an internal energy and the energy balance equation. In the general form, all kinds of energy may be considered but for this work only mechanical work and heat as sources of energy are accounted for in the balance equation. The energy balance equation states that the rate of the total energy $\mathcal{P}_{tot}$ is equal to the sum of the rate of external mechanical work $\mathcal{P}_{ext}$ done by forces acting on the body and of the flux of non-mechanical energy $\mathcal{P}_{th}$ due to heat or other sources

$$\mathcal{P}_{tot} = \mathcal{P}_{ext} + \mathcal{P}_{th} \quad . \tag{2.74}$$

The rate of the total energy is additively composed of the rate of the kinetic energy $\mathcal{K}$ and of the rate of the internal energy $\mathcal{E}$

$$\mathcal{P}_{tot} = \frac{\mathrm{d}}{\mathrm{d}t}\mathcal{K} + \frac{\mathrm{d}}{\mathrm{d}t}\mathcal{E} = \mathcal{P}_{kin} + \mathcal{P}_{int} \tag{2.75}$$

so that the balance equation reads

$$\mathcal{P}_{kin} + \mathcal{P}_{int} = \mathcal{P}_{th} + \mathcal{P}_{ext} \quad . \tag{2.76}$$

The respective terms in the spatial description are as follows

$$\mathcal{P}_{kin} = \frac{\mathrm{d}}{\mathrm{d}t}\mathcal{K} = \frac{\mathrm{d}}{\mathrm{d}t}\int_{\mathcal{B}_t} \frac{1}{2}\rho \boldsymbol{v}\cdot\boldsymbol{v}\ \mathrm{d}v = \int_{\mathcal{B}_t} \rho\dot{\boldsymbol{v}}\cdot\boldsymbol{v}\ \mathrm{d}v \tag{2.77}$$

$$\mathcal{P}_{int} = \frac{\mathrm{d}}{\mathrm{d}t}\mathcal{E} = \frac{\mathrm{d}}{\mathrm{d}t}\int_{\mathcal{B}_t} \rho e\ \mathrm{d}v = \int_{\mathcal{B}_t} \rho\dot{e}\ \mathrm{d}v \tag{2.78}$$

$$\mathcal{P}_{th} = \int_{\mathcal{B}_t} \rho r\ \mathrm{d}v - \int_{\partial\mathcal{B}_t} \boldsymbol{q}\cdot\boldsymbol{n}\ \mathrm{d}a = \int_{\mathcal{B}_t} (\rho r - \mathrm{div}[\boldsymbol{q}])\ \mathrm{d}v \tag{2.79}$$

$$\mathcal{P}_{ext} = \int_{\mathcal{B}_t} \boldsymbol{b}\cdot\boldsymbol{v}\ \mathrm{d}v + \int_{\partial\mathcal{B}_t} \boldsymbol{t}\cdot\boldsymbol{v}\ \mathrm{d}a \quad . \tag{2.80}$$

In (2.78), $\dot{e}$ is the rate of internal energy per unit mass and in (2.79) $r = r(\boldsymbol{x}, t)$ are heat sources per unit mass and $\boldsymbol{q} = \boldsymbol{q}(\boldsymbol{x}, t)$ is the heat flux per unit area. The second term of (2.80) may be reformulated by using the divergence theorem, the product rule and (2.24) which gives

$$\int_{\partial\mathcal{B}_t} \boldsymbol{t}\cdot\boldsymbol{v}\ \mathrm{d}a = \int_{\mathcal{B}_t} (\mathrm{div}[\boldsymbol{\sigma}]\cdot\boldsymbol{v} + \boldsymbol{\sigma}:\boldsymbol{d})\ \mathrm{d}v \tag{2.81}$$

where the relation $\boldsymbol{\sigma}:\boldsymbol{l} = \boldsymbol{\sigma}:\boldsymbol{d} + \boldsymbol{\sigma}:\boldsymbol{w} = \boldsymbol{\sigma}:\boldsymbol{d}$ is used and holds because of the skew symmetry of $\boldsymbol{w}$. Inserting all above relations in (2.76) one gets

$$\int_{\mathcal{B}_t} ([\rho\dot{\boldsymbol{v}} - \mathrm{div}[\boldsymbol{\sigma}] - \boldsymbol{b}]\cdot\boldsymbol{v} + \rho\dot{e})\ \mathrm{d}v = \int_{\mathcal{B}_t} (\boldsymbol{\sigma}:\boldsymbol{d} + \rho r - \mathrm{div}[\boldsymbol{q}])\ \mathrm{d}v \tag{2.82}$$

whereat the term in square brackets is identified as the linear momentum equation (2.68) and hence vanishes. Since the above equation must hold for all volumes the local form of the energy balance is given by

$$\rho \dot{e} = \boldsymbol{\sigma} : \boldsymbol{d} + \rho r - \operatorname{div}[\boldsymbol{q}] \quad . \tag{2.83}$$

For a purely mechanical isothermal problem, that means without heat flux ($\boldsymbol{q} = \boldsymbol{0}$) and heat sources ($r = 0$), it follows that the stress power $\boldsymbol{\sigma} : \boldsymbol{d}$ is equal to the rate of internal energy

$$\rho \dot{e} = \boldsymbol{\sigma} : \boldsymbol{d} \quad . \tag{2.84}$$

The stress power term is denoted as a power conjugate pair. That indicates that the stress is power conjugate to the rate of deformation. Further power conjugate pairs may be found by applying the pull-back operation which gives

$$J\boldsymbol{\sigma} : \boldsymbol{d} = \boldsymbol{\tau} : \boldsymbol{d} = \boldsymbol{P} : \dot{\boldsymbol{F}} = \boldsymbol{S} : \dot{\boldsymbol{E}} \quad . \tag{2.85}$$

2.4.5. Second law of thermodynamics

Since the first law of thermodynamics gives no information about the direction of energy transfer another principle is needed. For this purpose the intensive state variable entropy s per unit mass is introduced. The second law of thermodynamics postulates that the rate of change of the entropy is equal or greater than the rate of entropy input into the body. Or in other words, that the total production of entropy is greater than or equal to zero at all times. Mathematically, this is formulated by the Clausius-Duhem inequality in which T is the absolute temperature

$$\int_{\mathcal{B}_t} \rho \dot{s} \, \mathrm{d}v + \int_{\partial \mathcal{B}_t} \frac{\boldsymbol{q}}{T} \cdot \boldsymbol{n} \, \mathrm{d}a - \int_{\mathcal{B}_t} \rho \frac{r}{T} \, \mathrm{d}v \geq 0 \quad . \tag{2.86}$$

It may be transformed into a local form by using the divergence theorem

$$T \rho \dot{s} + \operatorname{div}[\boldsymbol{q}] - \rho r - \frac{1}{T} \boldsymbol{q} \cdot \operatorname{grad}[T] \geq 0 \quad . \tag{2.87}$$

With the local energy balance (2.83) one may rewrite the above equation as

$$\boldsymbol{\sigma} : \boldsymbol{d} - \rho(\dot{e} - T\dot{s}) - \frac{1}{T} \boldsymbol{q} \cdot \operatorname{grad}[T] \geq 0 \tag{2.88}$$

where the last term is identified as the heat conduction inequality

$$-\frac{1}{T}\boldsymbol{q}\cdot\mathrm{grad}[T]\geq 0 \tag{2.89}$$

which implies that heat flows against a temperature gradient. The first two terms of (2.88) represent the internal dissipation $\mathcal{D}_{int}$ and must be non-negative at all times as well. It is known as the Clausius-Planck inequality

$$\mathcal{D}_{int}=\boldsymbol{\sigma}:\boldsymbol{d}-\rho(\dot{e}-T\dot{s})\geq 0 \quad . \tag{2.90}$$

By applying the Legendre transformation it is possible to change the independent variables. The internal energy may be replaced by the Helmholtz free energy per unit mass Ψ as

$$\Psi=e-Ts \quad . \tag{2.91}$$

Applying the material time derivative of the free energy Ψ and inserting the result into (2.90) one gets the Clausius-Planck inequality in differential form with respect to the deformed unit volume

$$\mathcal{D}_{int}=\boldsymbol{\sigma}:\boldsymbol{d}-\rho(\dot{\Psi}+\dot{T}s)\geq 0 \tag{2.92}$$

and with (2.58) the inequality can be expressed with respect to the reference unit volume

$$\mathcal{D}_{int}=\boldsymbol{\tau}:\boldsymbol{d}-\rho_0(\dot{\Psi}+\dot{T}s)\geq 0 \quad . \tag{2.93}$$

In case of a purely isothermal process this inequality reduces to

$$\mathcal{D}_{int}=\boldsymbol{\tau}:\boldsymbol{d}-\rho_0\dot{\Psi}\geq 0 \quad . \tag{2.94}$$

2.5. Solution of the linear momentum balance with the finite element method

The finite element method (FEM) is a main tool in this work. Since the FEM gives the solution of an initial boundary value problem (BVP) of the momentum balance (2.68) it is possible to consider complex geometries in the computation. Developed constitutive models after implementation in a finite element code can be used for simulations. Thus, complicated material behavior may be analyzed and full processes are possible to compute, for instance the process chain of a particular product. The finite element method is briefly summarized in this section with both: the implicit and explicit solution strategy. For this purpose the initial BVP must be defined from which the weak form of the equation of motion is derived. This weak form is then discretized in space and linearized for using the Newton scheme as solution procedure. In the case of a dynamic problems the weak form must also be discretized in time.

2.5.1. Initial boundary value problem

An initial boundary value problem is defined by the equation of motion (2.68) and by boundary and initial conditions where the unknown displacement field is to be computed. This is in general not possible analytically and for this reason the FEM is used.
Two different boundary conditions are considered namely the displacement (Dirichlet) and the traction boundary condition (Neumann). For this purpose, the boundary is separated in two regions

$$\partial\mathcal{B} = \partial\mathcal{B}_u \cup \partial\mathcal{B}_\sigma \quad , \quad \partial\mathcal{B}_u \cap \partial\mathcal{B}_\sigma = \emptyset \tag{2.95}$$

with the boundary surface $\partial\mathcal{B}_u$ corresponding to the displacement boundary condition and the boundary surface $\partial\mathcal{B}_\sigma$ with respect to the traction boundary condition.
The initial boundary value problem with the respective boundary and initial conditions at $t = 0$ is given by

$$\begin{aligned} \mathrm{div}[\boldsymbol{\sigma}] + \boldsymbol{b} &= \rho\ddot{\boldsymbol{u}} \\ \boldsymbol{u} &= \overline{\boldsymbol{u}} \quad \text{on} \quad \partial\mathcal{B}_u \\ \boldsymbol{t} &= \overline{\boldsymbol{t}} \quad \text{on} \quad \partial\mathcal{B}_\sigma \\ \boldsymbol{u}(\boldsymbol{x},t)|_{t=0} &= \boldsymbol{u}_0(\boldsymbol{X}) \\ \dot{\boldsymbol{u}}(\boldsymbol{x},t)|_{t=0} &= \dot{\boldsymbol{u}}_0(\boldsymbol{X}) \quad . \end{aligned} \tag{2.96}$$

2.5.2. Weak form

The weak form of the equation of motion is obtained by multiplying the strong form of the equation of motion (2.68) with an arbitrary test function $\delta\boldsymbol{u} = \delta\boldsymbol{u}(\boldsymbol{x})$ which satisfies the geometric boundary conditions and integrating over the body

$$g(\boldsymbol{u}, \delta\boldsymbol{u}) = \int_{\mathcal{B}_t} (\mathrm{div}[\boldsymbol{\sigma}] - \boldsymbol{b} + \rho\ddot{\boldsymbol{u}}) \cdot \delta\boldsymbol{u} \; \mathrm{d}v = 0 \quad . \tag{2.97}$$

This approach is called the principle of virtual displacements by interpreting the function $\delta\boldsymbol{u}$ as a virtual displacement field. The only requirement is that this function must vanish at all points on the displacement boundary $\partial\mathcal{B}_u$. This requirement in combination with the divergence theorem and the product rule and inserting the traction boundary condition, the weak form results in

$$g(\boldsymbol{u},\delta\boldsymbol{u}) = \int\limits_{\mathcal{B}_t} [\boldsymbol{\sigma} : \mathrm{grad}[\delta\boldsymbol{u}] - (\boldsymbol{b} - \rho\ddot{\mathrm{u}}) \cdot \delta\boldsymbol{u}]\ \mathrm{d}v - \int\limits_{\partial\mathcal{B}_t^\sigma} \bar{\boldsymbol{t}} \cdot \delta\boldsymbol{u}\ \mathrm{d}a = 0 \quad . \tag{2.98}$$

The above equation is interpreted as the virtual work δW done by the virtual displacement where the integral

$$\delta W^{int} = \int\limits_{\mathcal{B}_t} \boldsymbol{\sigma} : \mathrm{grad}[\delta\boldsymbol{u}]\ \mathrm{d}v \tag{2.99}$$

is identified as the internal virtual work. The integrals with the external loads

$$\delta W^{ext} = \int\limits_{\mathcal{B}_t} \boldsymbol{b} \cdot \delta\boldsymbol{u}\ \mathrm{d}v + \int\limits_{\partial\mathcal{B}_t^\sigma} \bar{\boldsymbol{t}} \cdot \delta\boldsymbol{u}\ \mathrm{d}a \tag{2.100}$$

is the external virtual work and the integral of the inertia term times the virtual displacement

$$\delta W^{kin} = \int\limits_{\mathcal{B}_t} \rho\ddot{\mathrm{u}} \cdot \delta\boldsymbol{u}\ \mathrm{d}v \tag{2.101}$$

the kinetic virtual work.

2.5.3. Linearization

In case of a static analysis the inertia term in (2.98) is neglected so that the weak form of the equation of motion reduces to

$$g(\boldsymbol{u},\delta\boldsymbol{u}) = \int\limits_{\mathcal{B}_t} [\boldsymbol{\sigma} : \mathrm{grad}[\delta\boldsymbol{u}] - \boldsymbol{b} \cdot \delta\boldsymbol{u}]\ \mathrm{d}v - \int\limits_{\partial\mathcal{B}_t^\sigma} \bar{\boldsymbol{t}} \cdot \delta\boldsymbol{u}\ \mathrm{d}a = 0 \quad . \tag{2.102}$$

In general, this equation is non-linear due to material properties and the description of the geometry. An often used solution procedure is Newton's method to solve this kind of equation iteratively. Therefore, a first-order Taylor expansion at a known state is deployed.

However, the first-order Taylor expansion of (2.102) at the known displacement field state $\boldsymbol{u}$ yields

$$g(\boldsymbol{u}+\Delta\boldsymbol{u},\delta\boldsymbol{u}) = g(\boldsymbol{u},\delta\boldsymbol{u}) + \Delta g(\boldsymbol{u},\delta\boldsymbol{u}) = 0 \tag{2.103}$$

and is known as the linearization of the weak form of the equation of motion where Δg is the directional derivative of g in the direction of the increment of the displacement $\Delta\boldsymbol{u}$. To perform this derivative the integral and the spatial quantities must be transformed to the reference configuration where the directional derivative of G - the corresponding quantity in the reference configuration - may be carried out as follows:

$$\Delta G(\boldsymbol{u}+\Delta\boldsymbol{u}) = \frac{\mathrm{d}}{\mathrm{d}\varepsilon}[G(\boldsymbol{u}+\varepsilon\Delta\boldsymbol{u})]\bigg|_{\varepsilon=0} = DG\cdot\Delta\boldsymbol{u} \tag{2.104}$$

Assuming dead loads for $\boldsymbol{b}$ and $\overline{\boldsymbol{t}}$, the internal virtual work must be linearized only which gives rise to the following directional derivative

$$\begin{aligned} D\,\delta W^{int}\cdot\Delta\boldsymbol{u} &= D\Bigg[\int_{\mathcal{B}_t}\boldsymbol{\sigma}:\mathrm{grad}[\delta\boldsymbol{u}]\,\mathrm{d}v\Bigg]\cdot\Delta\boldsymbol{u} = D\Bigg[\int_{\mathcal{B}_0}\boldsymbol{\tau}:\mathrm{grad}[\delta\boldsymbol{u}]\,\mathrm{d}V\Bigg]\cdot\Delta\boldsymbol{u} \\ &= \int_{\mathcal{B}_0}\Big(D\boldsymbol{\tau}\cdot\Delta\boldsymbol{u}:\mathrm{grad}[\delta\boldsymbol{u}] + \boldsymbol{\tau}:D\,\mathrm{grad}[\delta\boldsymbol{u}]\cdot\Delta\boldsymbol{u}\Big)\,\mathrm{d}V \quad . \end{aligned} \tag{2.105}$$

In (2.105) two directional derivatives must be performed: that of the Kirchhoff stress and that of the gradient of virtual displacement. Applying (2.104) for the Kirchhoff stress, first it is pulled-back to the reference configuration and the product rule is adopted which gives

$$\begin{aligned} D\boldsymbol{\tau}\cdot\Delta\boldsymbol{u} &= D(\boldsymbol{F}\boldsymbol{S}\boldsymbol{F}^T)\cdot\Delta\boldsymbol{u} \\ &= D\boldsymbol{F}\cdot\Delta\boldsymbol{u}\boldsymbol{S}\boldsymbol{F}^T + \boldsymbol{F}D\boldsymbol{S}\cdot\Delta\boldsymbol{u}\boldsymbol{F}^T + \boldsymbol{F}\boldsymbol{S}D\boldsymbol{F}^T\cdot\Delta\boldsymbol{u} \quad . \end{aligned} \tag{2.106}$$

For a compact notation the incremental spatial velocity gradient $\Delta\boldsymbol{l}$ in terms of $\Delta\boldsymbol{u}$ is introduced as

$$\Delta\boldsymbol{l} = D\boldsymbol{F}\cdot\Delta\boldsymbol{u}\boldsymbol{F}^{-1} = \mathrm{grad}[\Delta\boldsymbol{u}] = \mathrm{Grad}[\Delta\boldsymbol{u}]\boldsymbol{F}^{-1} = \Delta\boldsymbol{F}\boldsymbol{F}^{-1} \tag{2.107}$$

from which the symmetric and skew-symmetric incremental tensors follow

$$\Delta\boldsymbol{d} = \frac{1}{2}(\Delta\boldsymbol{l}+\Delta\boldsymbol{l}^T) \quad , \quad \Delta\boldsymbol{w} = \frac{1}{2}(\Delta\boldsymbol{l}-\Delta\boldsymbol{l}^T) \quad . \tag{2.108}$$

Furthermore, the directional derivative of $\boldsymbol{S}$ may be rewritten in terms of $\Delta \boldsymbol{d}$ (cf. 2.43 and 2.51) which leads to

$$D\boldsymbol{S} \cdot \Delta \boldsymbol{u} = \frac{\partial \boldsymbol{S}}{\partial \boldsymbol{C}} : D\boldsymbol{C} \cdot \Delta \boldsymbol{u} = \mathbb{C} : \boldsymbol{F}^T \Delta \boldsymbol{d} \boldsymbol{F} \quad . \tag{2.109}$$

Inserting the relations (2.107) and (2.109) in (2.106), the linearization of the Kirchhoff stress is given in terms of spatial quantities by

$$D\boldsymbol{\tau} \cdot \Delta \boldsymbol{u} = \Delta \boldsymbol{\tau} = \mathbb{c} : \Delta \boldsymbol{d} + \Delta \boldsymbol{l} \boldsymbol{\tau} + \boldsymbol{\tau} \Delta \boldsymbol{l}^T \quad . \tag{2.110}$$

where the push-forward (2.53) of the material tangent modulus tensor $\mathbb{C}$ to the spatial tangent modulus tensor $\mathbb{c}$ has been employed.
The second directional derivative of (2.105), is that of the gradient of the virtual displacement and must be pulled-back first as well. By doing so and using the relation

$$\Delta \boldsymbol{F}^{-1} = D\boldsymbol{F}^{-1} \cdot \Delta \boldsymbol{u} = -\boldsymbol{F}^{-1} \text{grad}[\Delta \boldsymbol{u}] \tag{2.111}$$

and (2.107) to get the result in desired spatial terms, the directional derivative reads

$$\begin{aligned} D\text{grad}[\delta \boldsymbol{u}] \cdot \Delta \boldsymbol{u} &= D(\text{Grad}[\delta \boldsymbol{u}] \boldsymbol{F}^{-1}) \cdot \Delta \boldsymbol{u} = \text{Grad}[\delta \boldsymbol{u}] \; D\boldsymbol{F}^{-1} \cdot \Delta \boldsymbol{u} \\ &= -\text{Grad}[\delta \boldsymbol{u}] \boldsymbol{F}^{-1} \text{grad}[\Delta \boldsymbol{u}] = -\text{grad}[\delta \boldsymbol{u}] \Delta \boldsymbol{l} \quad . \end{aligned} \tag{2.112}$$

Inserting the results (2.110) and (2.112) in (2.105) and transforming the integral to the current configuration, the directional derivative needed for the linearization of (2.102) is completed and reads as follows

$$D\delta W^{int} \cdot \Delta \boldsymbol{u} = \int_{\mathcal{B}_t} (\frac{1}{J} \mathbb{c} : \Delta \boldsymbol{d} + \Delta \boldsymbol{l} \boldsymbol{\sigma}) : \text{grad}[\delta \boldsymbol{u}] \, \mathrm{d}v \quad . \tag{2.113}$$

2.5.4. Discretization in space

To solve the weak form of the equation of motion (2.98) the finite element method is used. Thereby, the body in the reference configuration $\mathcal{B}_0$ is subdivided in n_e non overlapping parts, the finite elements $\mathcal{B}_0^e$. The so-discretized body $\mathcal{B}_h^e$ serves as an approximation of the real body and consists of all finite elements

$$\mathcal{B}_0 \approx \mathcal{B}_0^h = \bigcup_{e=1}^{n_e} \mathcal{B}_0^e \quad . \tag{2.114}$$

A finite element is arranged by discrete nodes with specific degrees of freedom $\boldsymbol{u}_I$, here the displacements. The position of the I-th node is specified by the position vector $\boldsymbol{X}_I$ in the reference configuration or by $\boldsymbol{x}_I$ in the current configuration. The displacement and the geometry must be interpolated to solve for the new displacements. Within the framework of the isoparametric concept the same appropriate shape functions N_I are used for the approximation of the displacements and geometry. For an element with n nodes this approximation is given by summing over the product of the shape function at node I and the I-th nodal displacement or position vector, respectively:

$$\boldsymbol{u} \approx \boldsymbol{u}_e^h = \sum_{I=1}^{n} N_I \boldsymbol{u}_I \quad , \quad \boldsymbol{X}_e^h = \sum_{I=1}^{n} N_I \boldsymbol{X}_I \quad , \quad \boldsymbol{x}_e^h = \sum_{I=1}^{n} N_I \boldsymbol{x}_I \quad . \tag{2.115}$$

The virtual displacements are interpolated in the same manner and their gradient is given by the partial derivative of N_I only

$$\delta \boldsymbol{u}^h = \sum_{I=1}^{n} N_I \delta \boldsymbol{u}_I \quad , \quad \operatorname{grad}[\delta \boldsymbol{u}^h] = \sum_{I=1}^{n} \boldsymbol{B}_I \delta \boldsymbol{u}_I \tag{2.116}$$

where the matrix $\boldsymbol{B}$ includes the derivatives of the shape functions with respect to the spatial coordinates. Inserting these approximations in (2.98) ,the semi-discrete equation of motion is obtained

$$\delta \boldsymbol{u}^T (\boldsymbol{M} \ddot{\boldsymbol{u}} + \boldsymbol{f}^{int} - \boldsymbol{f}^{ext}) = \boldsymbol{0} \quad . \tag{2.117}$$

In this equation, the mass matrix $\boldsymbol{M}$, the nodal accelerations $\ddot{\boldsymbol{u}}$ and the internal and external nodal forces $\boldsymbol{f}^{int}$ and $\boldsymbol{f}^{ext}$ are already assembled. In detail, the respective terms are

$$\delta \boldsymbol{u}^T \boldsymbol{M} \ddot{\boldsymbol{u}} = \mathop{\mathsf{A}}_{e=1}^{n_e} \sum_{I=1}^{n} \sum_{K=1}^{n} \delta \boldsymbol{u}_I^T \boldsymbol{M}_{IK} \ddot{\boldsymbol{u}}_K \quad \text{with} \quad \boldsymbol{M}_{IK} = \boldsymbol{1} \int_{\mathcal{B}_t^e} \rho^e N_I N_K \, \mathrm{d}v^e \tag{2.118}$$

$$\delta \boldsymbol{u}^T \boldsymbol{f}^{int} = \mathop{\mathsf{A}}_{e=1}^{n_e} \sum_{I=1}^{n} \delta \boldsymbol{u}_I^T \boldsymbol{f}_I^{int} \quad \text{with} \quad \boldsymbol{f}_I^{int} = \int_{\mathcal{B}_t^e} \boldsymbol{B}_I^T \boldsymbol{\sigma}^e \, \mathrm{d}v^e \tag{2.119}$$

$$\delta \boldsymbol{u}^T \boldsymbol{f}^{ext} = \mathop{\mathsf{A}}_{e=1}^{n_e} \sum_{I=1}^{n} \delta \boldsymbol{u}_I^T \boldsymbol{f}_I^{ext} \quad \text{with} \quad \boldsymbol{f}_I^{ext} = \int_{\mathcal{B}_t^e} N_I \boldsymbol{b} \, \mathrm{d}v^e + \int_{\partial \mathcal{B}_t^{\sigma e}} N_I \bar{\boldsymbol{t}} \, \mathrm{d}a^e \tag{2.120}$$

where A is the assembling operator which sorts the contribution of each element to the global system of equations to the right position in this system. Due to the arbitrariness of the virtual nodal displacements, (2.117) reduces to a system of non-linear differential equations

$$\boldsymbol{M}\ddot{\boldsymbol{u}} + \boldsymbol{f}^{int} = \boldsymbol{f}^{ext} \quad . \tag{2.121}$$

In the static case, the inertia term in (2.121) is neglected and the resulting non-linear algebraic system of equations is solved with Newton's method. Therefore, the linearization of the variation of the internal work (2.103) must be discretized, too. Inserting the above approximations in (2.113) the discretized directional derivative for the linearization of (2.103)

$$D\delta W^{int} \cdot \Delta \boldsymbol{u} = \mathop{\mathsf{A}}_{e=1}^{n_e} \sum_{I=1}^{n} \sum_{K=1}^{n} \delta \boldsymbol{u}_I^T \boldsymbol{K}_{IK} \Delta \boldsymbol{u}_K \tag{2.122}$$

is obtained. The element tangent stiffness matrix is given by

$$\boldsymbol{K}_{IK} = \int_{\mathcal{B}_t^e} \left[\nabla N_I^T \boldsymbol{\sigma}^e \nabla N_K + \boldsymbol{B}_I^T \boldsymbol{D} \boldsymbol{B}_K \right] \mathrm{d}v^e \tag{2.123}$$

with the matrix form $\boldsymbol{D}$ of the spatial tangent moduli tensor $\mathbb{c}$ introduced in (2.110) and the unknown incremental nodal displacements $\Delta \boldsymbol{u}_I$. These displacements are computed by arranging the resulting linear system of equations from (2.103) and by inserting the internal and external nodal forces and the global tangent matrix $\boldsymbol{K}$. This leads to

$$\boldsymbol{K} \Delta \boldsymbol{u} = \boldsymbol{f}^{ext} - \boldsymbol{f}^{int} \tag{2.124}$$

where use is made of the arbitrariness of the virtual nodal displacements again. The nodal displacements are then updated $\boldsymbol{u}_{n+1} = \boldsymbol{u}_n + \Delta \boldsymbol{u}_I$ once the iterative scheme has converged.

The integrals appearing above are evaluated in a parametric space of a reference element (isoparametric concept). The necessary transformation of the integrals into the parametric space leads to a rational function. Thus, the integral is integrated numerically, for example by a Gaussian quadrature rule.

2.5.5. Discretization in time

The semi-discrete equation of motion (2.121) is discrete in space but continuous in time. To integrate this system of differential equations over time it is necessary to discretize it in time as well. Hence, a function is evaluated and solved at discrete times t_n. The time between two time points is the time step size Δt. The equations of motion discretized in space and time with the unknown displacements, velocities and accelerations at time $t_{n+1} = t_n + \Delta t_{n+1/2}$ read

$$\boldsymbol{M}\ddot{\boldsymbol{u}}_{n+1} + \boldsymbol{f}^{int}_{n+1} = \boldsymbol{f}^{ext}_{n+1} \tag{2.125}$$

assuming that all quantities are known at t_n. It is sometimes appropriate to introduce a damping matrix $\boldsymbol{C}$ to incorporate some amount of artificial energy dissipation proportional to the velocity. Then, the above equation reads

$$\boldsymbol{M}\ddot{\boldsymbol{u}}_{n+1} + \boldsymbol{C}\dot{\boldsymbol{u}}_{n+1} + \boldsymbol{f}^{int}_{n+1} = \boldsymbol{f}^{ext}_{n+1} \tag{2.126}$$

and may be integrated by an implicit or explicit scheme.

A popular explicit time integration scheme is the central difference method. Applying this rule for the acceleration at time t_n with $\Delta t_n = t_{n+1/2} - t_{n-1/2}$

$$\ddot{\boldsymbol{u}}_n = \frac{\dot{\boldsymbol{u}}_{n+1/2} - \dot{\boldsymbol{u}}_{n-1/2}}{\Delta t_n} \tag{2.127}$$

the velocity may be computed at time $t_{n+1/2}$

$$\dot{\boldsymbol{u}}_{n+1/2} = \dot{\boldsymbol{u}}_{n-1/2} + \Delta t_n \ddot{\boldsymbol{u}}_n \quad . \tag{2.128}$$

By applying the central difference approximation for the velocity, the unknown displacement at the time t_{n+1} is obtained

$$\dot{\boldsymbol{u}}_{n+1} = \frac{\boldsymbol{u}_{n+1} - \boldsymbol{u}_n}{\Delta t_{n+1/2}} \quad , \quad \boldsymbol{u}_{n+1} = \boldsymbol{u}_n + \Delta t_{n+1/2}\dot{\boldsymbol{u}}_{n+1/2} \quad . \tag{2.129}$$

The acceleration at time t_n appearing in (2.128) is given by the equation of motion (2.126) at time t_n

$$\ddot{\boldsymbol{u}}_n = \boldsymbol{M}^{-1}(\boldsymbol{f}^{ext}_n - \boldsymbol{f}^{int}_n - \boldsymbol{C}\dot{\boldsymbol{u}}_{n-1/2}) \tag{2.130}$$

where the inverse of the mass matrix has to be computed. This is much more efficient if this matrix is lumped. So, only vector operations must be performed. At this point, all other unknowns, like the nodal force vectors and the stresses at the time t_{n+1} are computable. Since this integration scheme is explicit -all unknown quantities at the end of the time step depend on known quantities at the beginning of the step- it is subjected to requirements on the size of the time step to ensure stability of the scheme. This so-called critical time step Δt_{crit} may be estimated by the maximum eigenfrequency of all elements

$$\Delta t_{crit} = \frac{2}{\max\limits_e \omega^e} = \min_e \frac{l^e}{c^e_d} \tag{2.131}$$

with the characteristic element dimension l^e and the dilatational wave speed c_d^e of the material specified for an element.

3. Mechanical behavior of amorphous thermoplastic polymers

In the previous chapter, the continuum mechanical basics and the solution of an initial boundary value problem were depicted without giving a constitutive relation between deformation and stress. The present chapter deals with the physics of amorphous thermoplastic polymers relevant for this work and the resulting mechanical behavior. Additionally, uniaxial tensile tests will be presented which are performed in the range from room temperature till above the glass transition at different strain rates. Two amorphous thermoplastic polymers, the common and widely used PMMA (polymethyl methacrylate) and PC (polycarbonate) are examined. The literature along with own experimental findings give indications for a "feature list" for the constitutive model which will be deployed in the next chapter.

3.1. Amorphous thermoplastic polymers – general foundations

In this section, the molecular structure of amorphous thermoplastic polymers is described briefly at first. Following this, the mechanical behavior is depicted deploying the example of uniaxial tensile tests. Especially, the characteristics of yielding and the large strain behavior is discussed, both in the context of applied strain rate and temperature. The above mentioned polymer physics may be found in the works of Haward and Young (1997), Strobl (2007), Ward and Sweeney (2005), Rösler et al. (2003), Treloar (1974), G'Sell et al. (1992), Boyce et al. (1988) are used for this section particularly.

3.1.1. Molecular structure and entanglement network

Polymers consist of long macromolecular chains with thousands of repeating identical units (monomers) linked by covalent bonds. For example, the monomers of PMMA and PC are displayed in Fig. 3.1. The degree of polymerization n, defined as the average number of monomers, as well as the structure of the monomer itself highly influences the mechanical behavior of the polymer due to different possibilities of molecular motions.

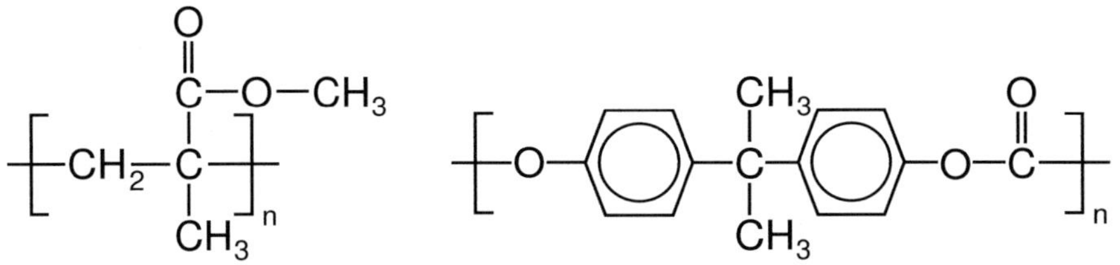

(a) polymethyl methacrylate (PMMA)

(b) polycarbonate (PC)

Figure 3.1.: Chemical structure of commonly used amorphous thermoplastic polymers

Polymers are distinguished by the amount of chemical cross-links which strongly influences the mechanical response. While thermoplastics are not chemically cross-linked at all, elastomers consist of a few and thermosets of a high amount of chemical cross-links. Both thermoplastic polymers, PMMA and PC, are almost completely amorphous due to their irregular chemical structure, caused by the large side groups attached to the backbone of the chain. In contrast to that, if the monomer consists of a regular structure the polymer tends to crystallize in certain regions but still contains amorphous parts. The polymer is then called semicrystalline; typical examples of semicrystalline thermoplastics are polypropylene (PP) and polyethylene (PE).

However, the microstructure of amorphous thermoplastics is understood as a network of randomly coiled macromolecules connected by physical cross-links (entanglements). This network is illustrated in Fig. 3.2a. The magnification illustrates the conformation of the chain due to random rotations of rigid chain segments about covalent bonds between atoms under the restriction of the valence bond angle. These intramolecular bonds inside the chain may be considered as rigid compared to the much weaker intermolecular van der Waals interactions between the molecular chains.

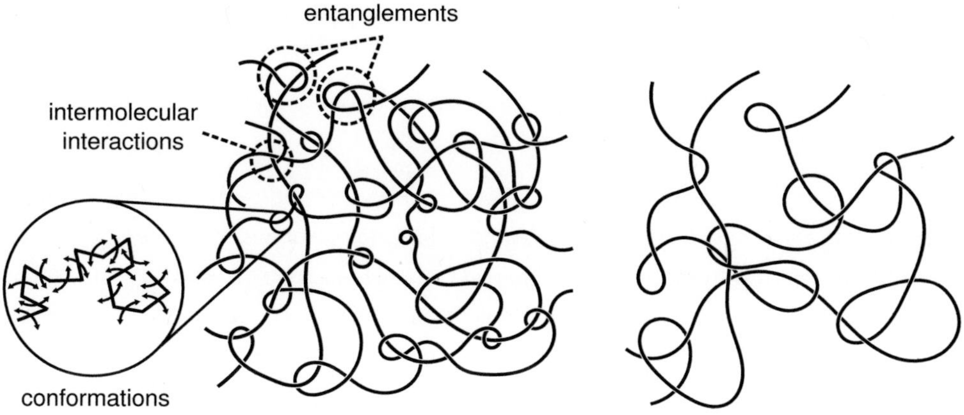

(a) below T_g (b) around T_g

Figure 3.2.: Network of entangled polymer chains

Thermal energy leads to an increased motion and mobility of chain segments and thus to a higher distance of the chains to each other. With further increasing temperature there is a rapid change in the distance, respectively, the specific volume $\nu = \rho^{-1}$ at which the volume increases more rapidly than at lower temperatures (Fig. 3.3). This is explained with the concept of free volume. The free volume is the empty space between polymer chains and increases at a specific temperature. This temperature at which this rapid change occurs is called the glass transition temperature T_g because the polymer changes from the glassy to the rubbery state [1].

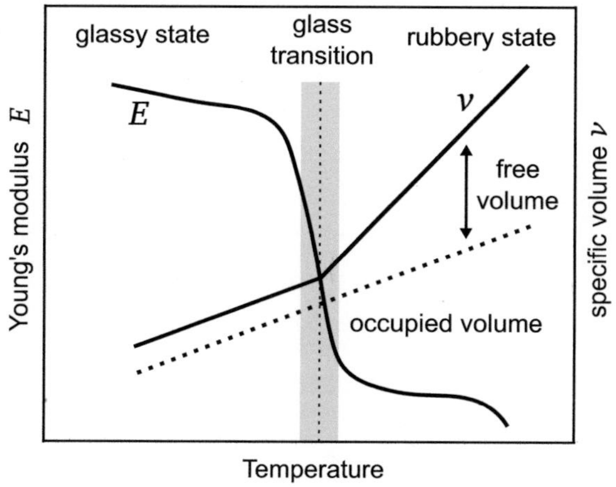

Figure 3.3.: Change of stiffness and specific volume at glass transition

[1] According to the measurement technique of the glass transition temperature an precise temperature is not measurable. Rather a glass transition region is identified.

In Fig. 3.2b, the molecular network at T_g is sketched. Obviously, the mass density ρ of the polymer is lower than below T_g (Fig. 3.2a) and more free volume is available. Furthermore, larger segments of the polymer chain become mobile at T_g. This gives rise to a relaxation process and is called the α-relaxation[2].

In the glassy state, i.e. at temperatures below T_g, the molecular chains are frozen in a state which is essentially a supercooled liquid. This microstructure is either isotropic due to a random orientation of the coiled chains or may be anisotropic due to processing or mechanical straining from which the chains orient and stretch. In Fig. 3.4 a molecular network stretched in the direction of applied stretch or stress is depicted. Based on this, it is obvious that the extensibility at high strains is dependent on the entanglement density since the entanglements prevent a further stretch of the chains.

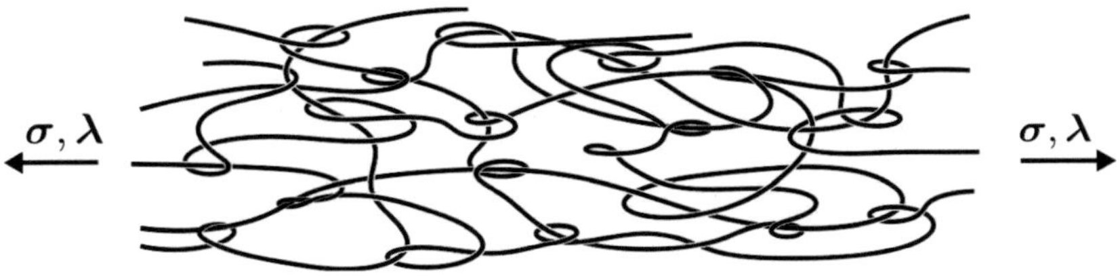

Figure 3.4.: Stretched and oriented polymer chains

The concept of an underlying entangled network of macromolecular chains was motivated by the observation that amorphous thermoplastics stretched in the glassy state tend to deform to their original shape if the temperature is raised above T_g (Haward and Young, 1997). This suggests that a number of permanent entanglements between the chains exist which are responsible for the *shape memory effect*. The experiment shown in Fig. 3.5 illustrates this phenomenon. An injection molded tensile test specimen made of PMMA (Fig. 3.5a) is heated at approximately glass transition temperature and is then twisted (Fig. 3.5b) before it is cooled to room temperature[3]. The annealing above T_g at 120 °C for 20 minutes leads to a complete redeformation to the original shape as well as a shrinkage of the specimen (Fig. 3.5c). Thereby, the deformation of the specimen takes place due to coiling of the chains to a preferred random isotropic state between the still existing entanglements. The additional shrinkage which becomes obvious when Figs. 3.5a and 3.5c are compared, results from the earlier molding process at which the

[2] With decreasing temperature the relaxation processes after the α-relaxation at T_g are called β-relaxation, at which in general side group motions occur, then γ-relaxation and so on.

[3] In Struik (1990) a far more detailed discussion about twisted cylindrical rods and the resulting force at deformation above T_g to the original shape may be found.

polymer chains were oriented and stretched before rapid cooling led to the anisotropic *frozen-in* microstructure.

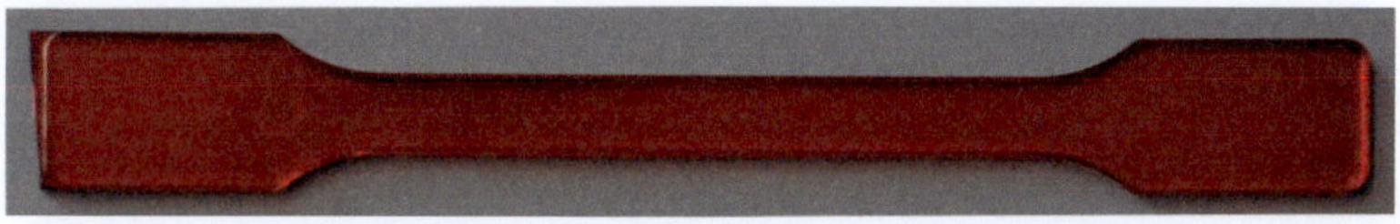

(a) initial state after processing (injection molding)

(b) twisting of locally heated region near glass transition temperature

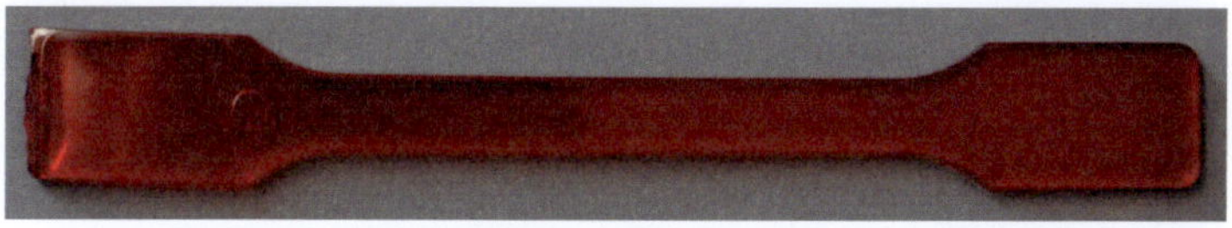

(c) deformation after annealing at 120 °C (T_g=105 °C) for 20 minutes

Figure 3.5.: Shrinkage and shape memory effect of an injection molded and twisted test specimen made of PMMA

3.1.2. Mechanical characteristics

The strain response of a glassy polymer subjected to small stresses is practically independent of the loading rate, reversible and proportional to the stress. When it is stressed further to few percent of strains it shows a pronounced viscoelastic behavior where the loading rate as well as the temperature have strong influences on the mechanical behavior. Thereby, a high strain rate corresponds to a low temperature and vice versa (Williams et al., 1955). The elastic mechanical behavior in the glassy state is mainly determined by the weak intermolecular interactions between the molecular chains. The intramolecular covalent bonds inside the chain are much stronger than the intermolecular interactions and thus less important for the mechanical response at small strains and may be considered as rigid. For example, this results in the typical low Young's modulus of thermoplastic polymers which is approximately hundred times lower than that of metals. During the loading, the atoms move from an energetic preferred state of equilibrium which is attained again at unloading; this is called energy elasticity. Different relaxation processes (rearrangement of molecules) take place with different relaxation times and activation energies. This determines the viscoelasticity of the material. Around the glass transition temperature the polymer enters the rubbery state which comes along with a drop of the stiffness (Fig. 3.3) and a more pronounced viscoelastic behavior. Above T_g, the molecules get stretched between the entanglements during the loading and relax to the original unstretched state after unloading. So, the elasticity in the rubbery state is determined by the entropy of the conformation of the macromolecule (entropy elasticity).

If the stress is further increased and the temperature is below the glass transition temperature the behavior becomes more non-linear and large plastic deformations may occur by shear yielding (in absence of brittle failure or crazing). The plastic deformation takes place essentially at constant volume and is characterized by a stress peak followed by intrinsic strain softening. The latter typically results in inhomogeneous deformation such as shear banding. In the subsequent plastic deformation process progressive hardening with increasing strain is observable till a limit stretch is reached and no further plastic deformation takes place. The material becomes elastic again until the specimen fails by rupture of covalent bonds. As stated before, the plastic deformation is reversible and the material returns to the original shape when the temperature is raised above the glass transition temperature. Thus, it is not a true plastic deformation of the material but rather a highly nonlinear viscoelastic behavior. The mechanical behavior of PC is depicted in Fig. 3.6 where tensile true stress-log. strain curves at different temperatures and strain rates are plotted (G'Sell et al., 1992) [4].

The onset of yielding is temperature, strain rate as well as pressure dependent, where the latter results in a higher yield stress in compression than in tension. The yield point is raised at lower temperatures and higher loading rates and respectively, reduced at higher temperatures and lower rates. A variation of strain rate causes a shifting of the whole curve while different temperatures not only shift the curves but also results in a decreasing hardening modulus with increasing temperatures (Fig. 3.6a,b). Thus, the plastic deformation behavior is strongly viscoplastic in nature. The origin of this behavior is associated with the mechanism of molecular chain segment rotations taking place once the applied stress is high enough to overcome the resistance of the intermolecular bonds. This is a thermally activated process and is time-dependent due to specific relaxation times. In the rubbery state[5] a difference between the elastic and plastic deformation is not specifiable and a yield point is not observable anymore (Fig. 3.6b). This suggests a

[4] The viscoelastic response is not observable due to the large strain scale plotted and due to the monotonic process (no hysteresis).

[5] The glass transition temperature of PC is around 145 °C.

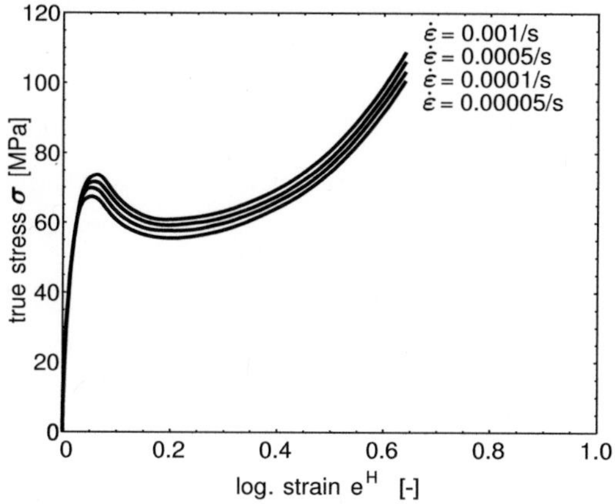

(a) Influence strain rates at a temperature of 25 °C

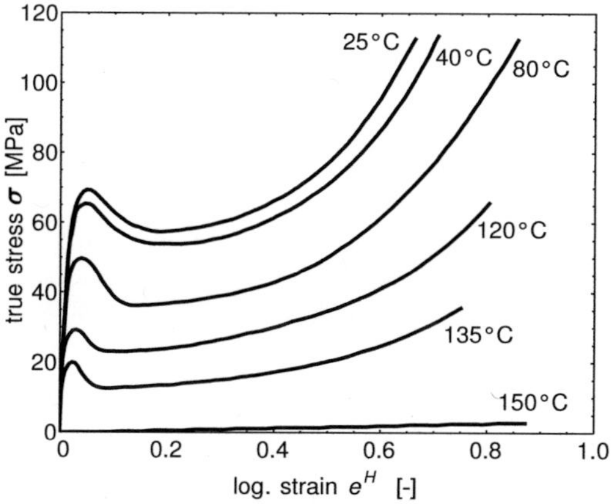

(b) Influence of temperature at a $\dot{\varepsilon} = 10^{-4}s^{-1}$

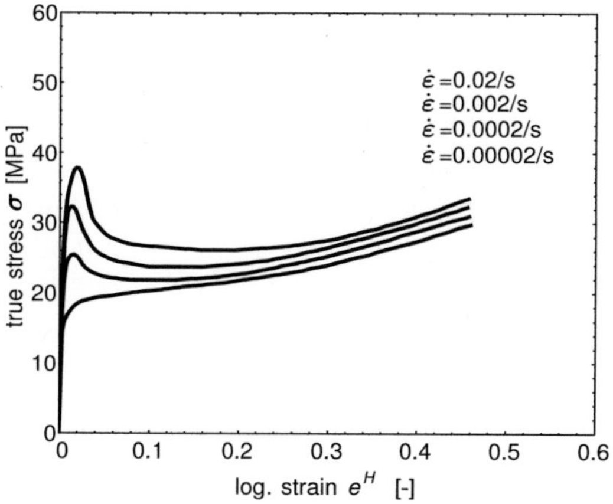

(c) Influence of strain rate at higher temperature (125 °C)

Figure 3.6.: Mechanical behavior of PC in a uniaxial tensile test (G'Sell et al., 1992) change from a solid to a liquid-like behavior at small strains. The resistance against chain segment rotations is strongly reduced due to a higher free volume available.

Instantly after the onset of yielding the stress drops and the polymer softens to a lower stress level (Fig. 3.6a). The amount of stress drop depends on the temperature and the strain rate as well (Fig. 3.6b,c). Furthermore, there is a dependence on the mechanical and thermal loading history:

- the softening vanishes at cyclic loading; a lower yield stress at reverse and re-loading is observed. Thus, thermoplastics exhibit a strong Bauschinger effect.
- the stress peak a the onset of yield is significantly higher for an annealed material than for a quenched material, yet the strain level after softening is the same in both cases.

Physical aging leads to a recovery of the initial yield stress (Boyce et al., 1988). The mechanism of softening on the microscale is still not very well understood and is assumed to be a result of an average restructuring of the molecular chains (Boyce et al., 1988). Around T_g a stress drop (softening) is not longer observable which, e.g. in tensile specimens, results in a much more homogeneous deformation.

During a loading beyond 10-20 % of strain the material begins to harden progressively with increasing strain. The hardening modulus decreases with increasing temperature (Fig. 3.6b) and shows a rate dependence at temperatures around T_g (see Sec. 3.2). Below T_g, the hardening results from the large stretching and orientation of the molecular chains between the entanglements in the direction of loading (Fig. 3.4). When the chains are fully stretched the limit stretch is reached and the covalent bonds are loaded elastically. This leads to local hardening and overcompensates the necking. Thus, the necking area propagates over the whole specimen due to stretching the neighboring unstretched chains which exhibit a lower resistance against plastic deformation. The complete deformation behavior of a tensile test in the glassy state is depicted in Fig. 3.7 in which the shear band formation and the subsequent development and growth of the neck is observable.

The assumption of a pure entropic origin of the hardening as in the theory of rubber elasticity is in contrast to the wrong trend in temperature (decreasing hardening modulus at increasing temperature). Rather, it shows a viscoelastic behavior. The true origin is still an open question and is discussed in literature in the recent years from the side of experimental studies (Govaert and Tervoort (2004), Wendlandt et al. (2005), Chen and Schweizer (2009), Senden et al. (2010), Senden et al. (2012)) as well as from molecular dynamic simulations (Hoy and Robbins (2006), Hoy and Robbins (2008)). At temperatures around and above T_g the molecules are more and more able to slip at each other due to the higher free volume; the material behaves like a viscous fluid. A theory to explain the mechanism of viscous flow is the reptation of molecular chains (de Gennes (1983), Doi and Edwards (1978), de Gennes (1971)). It is assumed that at loading the chains may be pulled out of the network through a tube formed by entanglements of surrounding chains. The tube size may increase with increasing temperature due to increased free volume. In contrast to these high temperatures it is expected that the tube diffusion does not takes place in the glassy state.

The local strain rate and temperature increase in the necking area. In detail, a portion of the plastic work is stored in the stretch of the chains and a portion is dissipated and converted into heat as a result of molecular motions which alleviates further plastic deformation. For example, this leads to a local temperature rise of a maximum of 20 °C at a strain rate of $\dot{\varepsilon} = 0.1\text{s}^{-1}$ applied on PMMA (Arruda et al., 1995).

Figure 3.7.: Deformation behavior of PC in a uniaxial tensile test below T_g

The effect of aligned molecular chains due to stretching the material or due to processing such as injection molding on the mechanical behavior is mainly observable in the plastic region. It is negligible for the elastic behavior (Arruda et al., 1993). The most important influences of molecular pre-orientation on the plastic response are the appearance of anisotropic yielding, the vanishing of softening and the raised hardening modulus and limited extensibility if loaded in the direction of the pre-stretch. In Fig. 3.8 true stress-log. strain curves of compression tests are depicted which show these effects for PC compressed to a certain amount of strain and reloaded in the direction of the pre-stretch and perpendicular to that direction (Arruda et al., 1993). Note, the direction of pre-stretch is not the loading direction at compression rather it is the direction perpendicular to the pre-stretch. When the material is reloaded in that direction the molecular network first deorients and then orients in the direction perpendicular of the loading direction which results in a lower yield stress. A reload of the material in direction of pre-stretch results in a higher yield stress because the network is already stretched in that direction and no further large stretch is possible.

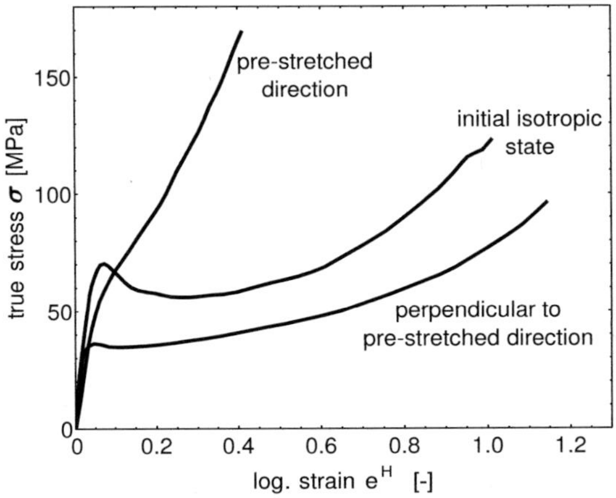

Figure 3.8.: Effect of pre-stretched PC on the inelastic behavior (uniaxial compression true stress - log. strain curves at strain rate of $\dot{\varepsilon} = 0.001\text{s}^{-1}$ and room temperature, pre-stretched to $e^H = -0.75$) (Arruda et al., 1993)

3.2. Experimental study

In order to obtain consistent material data for the mechanical behavior below and above the glass transition, in the present work uniaxial tensile tests are performed on PMMA and PC at three different strain rates and temperatures from room temperature till beyond the glass transition temperature of the specific material. To measure the local strain field, a video extensometer is used with a subsequent evaluation utilizing a digital image correlation (DIC) tool. The experimental setup, the evaluation procedure and the findings are presented in the following sections.

3.2.1. Test setup

The experimental setup consists of a testing machine, an oven, a load cell and a video extensometer with the recording and evaluating software.

The servohydraulic Instron 1342 testing machine was upgraded with the digital electronic controller Fasttrack 8800, the software Wavematrix by Instron (2012) and a 10 kN strain gauge load cell. The camera of the video extensometer for a non-contact optical measurement is placed close to the clamped test specimen to record it as planar as possible and is supported by diffuse light. This system, namely Q-400, of the company Limess works with a ccd gray scale image sensor, with a resolution of 2 MP and a maximum recording frequency of 15 Hz, to record the deformation of a test specimen. The recorded images are saved on a PC to evaluate the strain field afterwards with the correlation and sensor controlling software Istra 4D by Dantec Dynamics A/S (2012). The sensor is controlled with a PC and the timing hub TU-4XB which triggers the recording and synchronizes the images with the incoming force signal of the load cell. The test specimens are fixed in a self constructed clamping. In Fig. 3.9 the setup without the oven is shown.

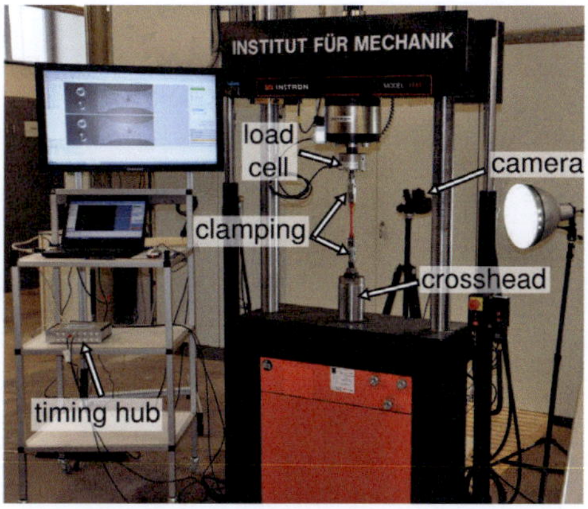

(a) Testing machine in the laboratory

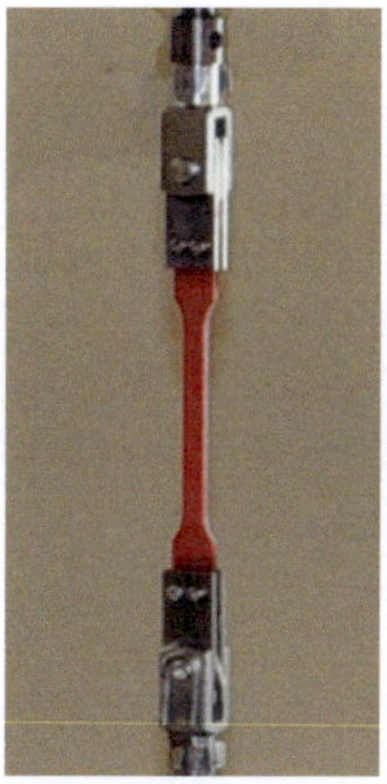

(b) Clamping of specimen

Figure 3.9.: Servohydraulic testing machine with measurement system

For the experiments at higher temperatures, an oven is installed around the extended crosshead (see Fig. 3.10). The maximum temperature is 350 °C which is adjusted by a controller of the company Eurotherm. To use the video extensometer in the case of these experiments as well, the test specimens are filmed through a glass window in the oven which is lighted from inside.

The specimens made of PMMA were injection molded and kindly provided by the Institute of Microstructure Technology (IMT), KIT. The sketch with all dimensions is given in Fig. 3.11a. Due to the molding, the surface is very smooth so that no further specimen treatment is necessary (Fig. 3.11b). Since the geometry of the PMMA test specimens is not optimal for clamping, the geometry for the PC test specimens is adopted from the

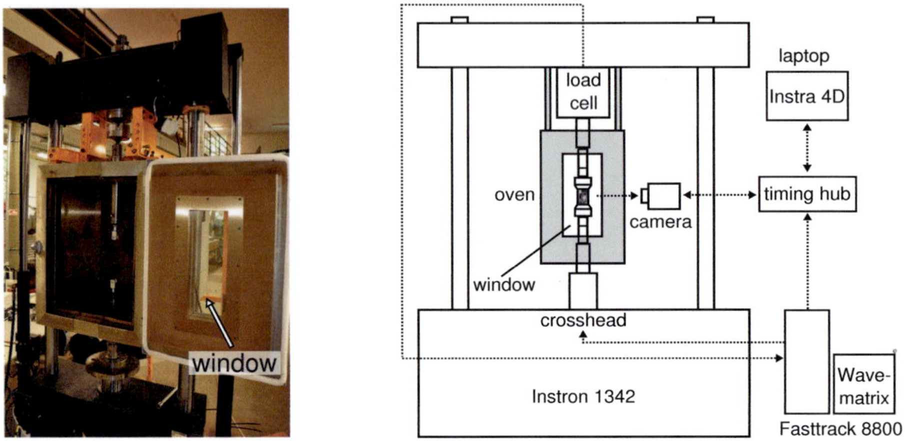

(a) Testing machine with oven (b) schematical experimental setup with oven

Figure 3.10.: Built-on oven at testing machine

work of Becker (2009) and is shown in Fig. 3.12a. These specimens are made by waterjet cutting from injection molded plates (Lexan by Sabic) and are kindly provided by the Deutsches Kunststoff Institut (DKI). After the cutting, the surface is very rough and a specimen preparation is necessary (Fig. 3.12b). The specimens are burred by hand with sand paper which gives a relative smooth surface (Fig. 3.12c). A rough surface would result in an early failure of the specimen due to scratches.

For the DIC based deformation measurement system explained in Sec. 3.2.2, the test specimens must have a random speckle pattern on its surface. Therefore, one surface of the test specimen is painted white followed by spraying black points on it which gives a random speckle pattern, see Fig. 3.11c and Fig. 3.12d.

3.2.2. Evaluation

The recorded images of the test specimen during the experiment are processed with the DIC software Istra 4D. The very basics of the digital image correlation method are briefly summarized in the following. A fundamental description may be found in Sutton et al. (2009) and Pan et al. (2009).

The idea of DIC is to compute the displacement and the strain field by comparing the recorded images of the incremental deformation steps by taking many points spread over the whole specimen into account and not, as in the method of G'Sell et al. (1992), only few in a row in the middle of the specimen which is less accurate. For this purpose,

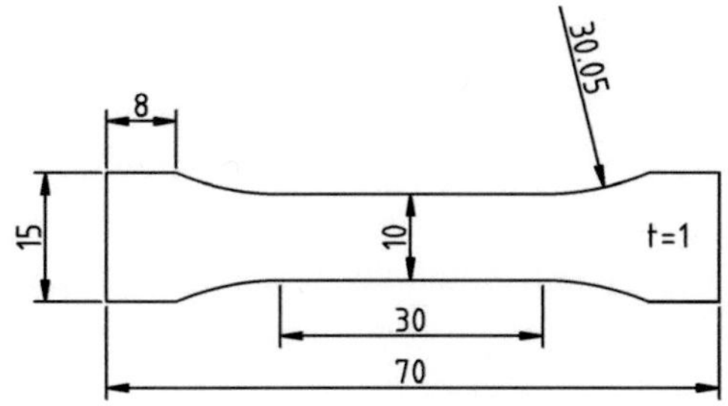

(a) sketch of specimen

(b) specimen after moulding

(c) speckle pattern

Figure 3.11.: Test specimen of PMMA (thickness 1mm)

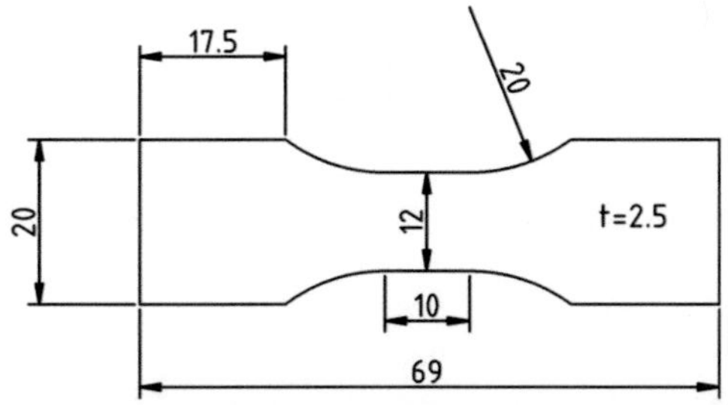

(a) sketch of specimen

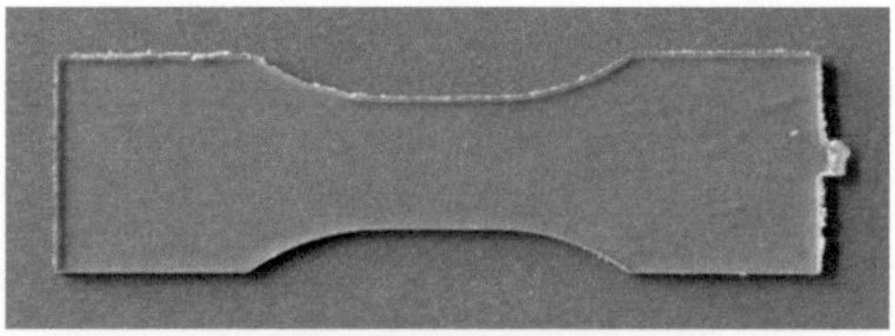

(b) specimen after cutting

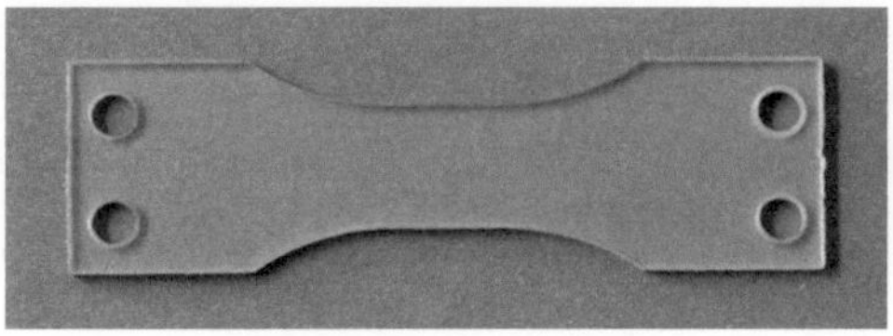

(c) specimen after burring

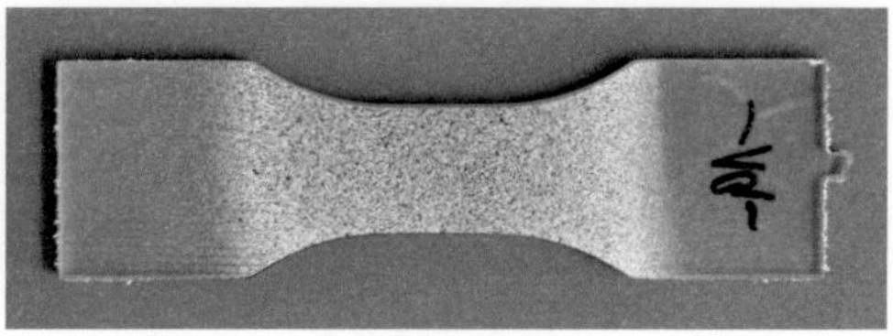

(d) speckle pattern

Figure 3.12.: Test specimen of PC (thickness 2.5mm)

a region in the reference image must be defined where the correlation process should take place. This region is then subdivided in a grid where in the following at each point the displacement is computed. Around the grid points, a squared reference subset of $(2M+1) \times (2M+1)$ pixels is defined with the reference center point $P(x_0, y_0)$ of the specific subset. Instead of a single pixel, a subset is chosen for a better matching of the reference center point P and the point $P'(x'_0, y'_0)$ in the deformed or target subset. This works best if the gray scale pixel are randomly distributed like the speckle pattern described above. Increasing the size of the subset increases the accuracy of the method but increases the numerical effort as well.

The main task is now to find the center point P' of the deformed subset corresponding to the reference center point P. If this point is found, the displacement vector $\boldsymbol{u}$ can be computed as the differences of the positions of the reference center point and the center point of the deformed subsets as depicted in Fig. 3.13.

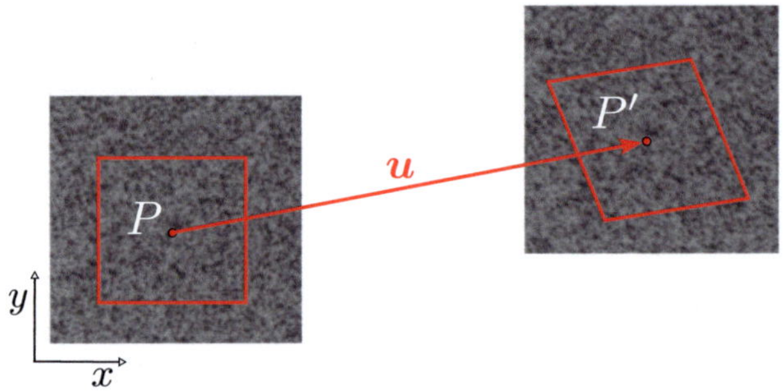

Figure 3.13.: Deformation of a reference subset (left) to the target subset (right) due to translation, rotation, stretch, shear and the corresponding displacement $\boldsymbol{u}$ of the subset center point P to P'

For this matching procedure of the points, it is necessary to consider the deformation of the subset, i.e. all points inside this subset. With the assumption that neighboring points in the reference subset remain neighbors in the deformed subset, an arbitrary point $Q(x_i, y_j)$ in the reference subset may be mapped to the point $Q'(x'_i, y'_j)$ in the deformed subset as

$$\left.\begin{aligned} x'_i &= x_i + \xi(x_i, y_j) \\ y'_i &= y_i + \eta(x_i, y_j) \end{aligned}\right\} \quad i, j = -M, ..., M \tag{3.1}$$

with the shape functions ξ and η. For example, first-order shape functions may consider translation, rotation, stretch and shear of the subset and yield

$$\xi_1(x_i, y_j) = u + \frac{\mathrm{d}u}{\mathrm{d}x}(x_i - x_0) + \frac{\mathrm{d}u}{\mathrm{d}y}(y_j - y_0) \tag{3.2}$$

$$\eta_1(x_i, y_j) = v + \frac{\mathrm{d}v}{\mathrm{d}x}(x_i - x_0) + \frac{\mathrm{d}v}{\mathrm{d}y}(y_j - y_0) \tag{3.3}$$

with the center point displacement u and v in the x- and y- direction. While zero-order shape functions would result in mapping of rigid body motions only, second-order functions are able to consider the curvature of the subset boundaries. Exemplary, the deformation of a subset for first-order shape functions is shown in Fig. 3.13. In a uniaxial tensile test, basically translation and stretch of a subset takes place. This situation is depicted in Fig. 3.14. Compared to translation and stretch, a small amount of local rotation and shear occur as well in case of shear banding or anisotropic materials.

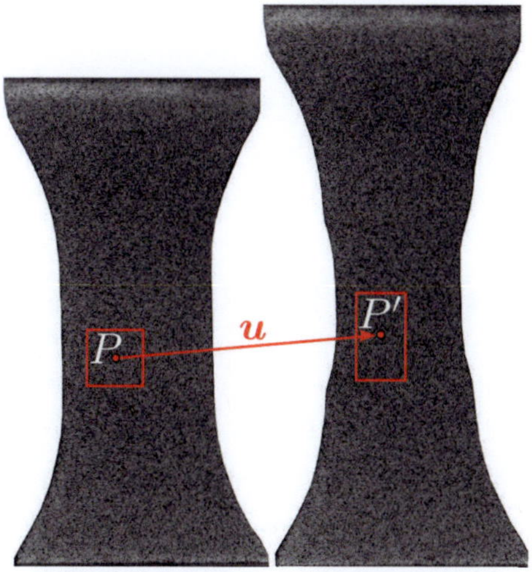

Figure 3.14.: Deformation of a subset in a tensile test: translation and stretch

The similarity of the gray scale values in the reference subset and the deformed target subset may be computed with different correlation criteria. The simplest one is the sum of squared differences (SSD)

$$C(u, v, \frac{\mathrm{d}u}{\mathrm{d}x}, \frac{\mathrm{d}u}{\mathrm{d}y}, \frac{\mathrm{d}v}{\mathrm{d}x}, \frac{\mathrm{d}v}{\mathrm{d}y}) = \sum_{i=-M}^{M} \sum_{j=-M}^{M} [f(x_i, y_j) - g(x'_i, y'_j)]^2 \tag{3.4}$$

with the gray scale value $f(x_i, y_j)$ at point (x_i, y_j) of the reference subset and the gray scale value $g(x_i', y_j')$ at point (x_i', y_j') of the target subset, respectively. The value of the correlation criterion C would be never equal to zero in reality. Therefore, a critical limit value must be set at which the calculation is stopped if the maximum of similarity is found. At this point the displacement vector can be computed. Several algorithms exist to enhance this computation. One of them is the photometric mapping which considers influences of light changes during the recording and hence improves the correlation accuracy.

If the displacement vector of all grid points is known, the strain field may be calculated. For this purpose a function for the displacement is needed to calculate the strain by differentiation of the displacement. For example, this can be done by a point wise fitting technique using a polynomial function. The assumption of a linear displacement field yields

$$u(i,j) = u_0 + u_1 x + u_2 y \tag{3.5}$$

$$v(i,j) = v_0 + v_1 x + v_2 y \tag{3.6}$$

with the unknown coefficients $u_{i=0,1,2}$ and $v_{i=0,1,2}$ and the known displacements $u(i,j)$ and $v(i,j)$ obtained by the DIC. The indices $i, j = -m, .., m$ are local coordinates in a strain calculation area of $(2m+1) \times (2m+1)$ discrete points where the displacement is known. The polynomial coefficients may be determined by a least-square method and the desired strain field is simply gained by differentiation of (3.5) and (3.6) with respect to the coordinates x and y. For example, the xx-component of the Green-Lagrange strain tensor (2.16) is given by $E_{xx} = u_1 + \frac{1}{2}(u_1^2 + v_1^2)$. This strain tensor is computed by the software Istra 4D. In Fig. 3.15, the E_{xx} (loading direction) strain field at different states of deformation is shown for the PC specimen. In Fig. 3.15a, the region which is used for the correlation is indicated. It is sufficient to restrict the evaluation to the rectangular center region of the specimen where the largest deformation takes place. This is clearly observable in Fig. 3.15b-d. It is noted that an out-of-plane displacement of the specimens is to be avoided to guarantee the best possible accuracy of the strain measurement system.

The aim of this experimental work is to measure the local strain field needed to compute the true cross section of the specimen to obtain the real mechanical material behavior in terms of stress-strain response. Thus, the strain is taken from a small region only, analogously to a small strain gauge. In this gauge, the strain field obtained from the DIC is averaged. The question is how to choose the size of the strain averaging region in order

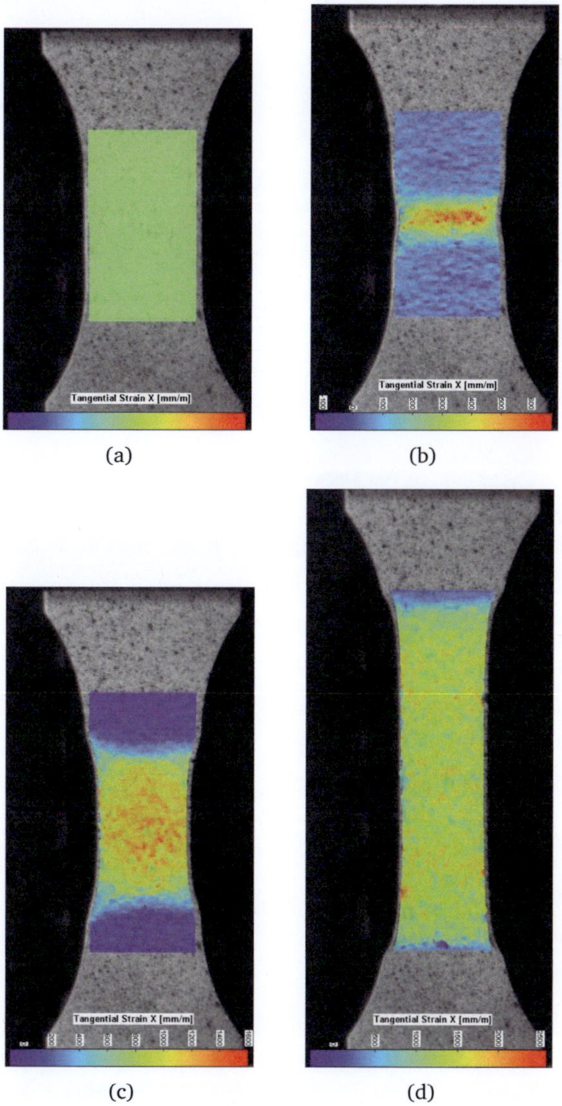

Figure 3.15.: Strain field in loading direction at different deformation states of PC

to get the most accurate response. To answer this question, different sizes of rectangular strain gauges are used in the region where maximum strains occur. The size differs from 1 pixel up to 30 pixel in loading direction. Perpendicular to the loading direction, almost the whole width of the specimen is chosen. In Fig. 3.16, these regions are sketched on top of the strain field of Fig. 3.15b.

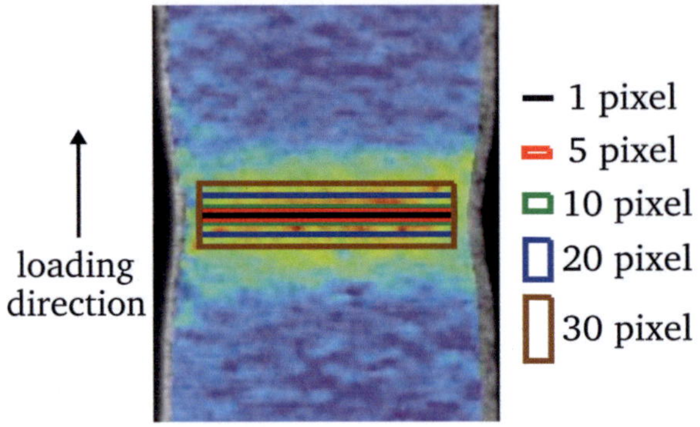

Figure 3.16.: Different sizes of strain gauges

The selection of the size of the gauge strongly influences the results of the strain measurement (Fig. 3.17). Clear differences are observable if the size is larger than 5 pixel. Almost no difference can be seen between a size of 5 pixel and 1 pixel. The larger the size the more an averaged strain field is obtained and the less accurate the measurement is. On the basis of this result, in the present work a size of about 5 pixel for the gauge is used in the evaluations to accurately capture the local strain field. In the work of Kotlik (2011), the same size was found to be appropriate.

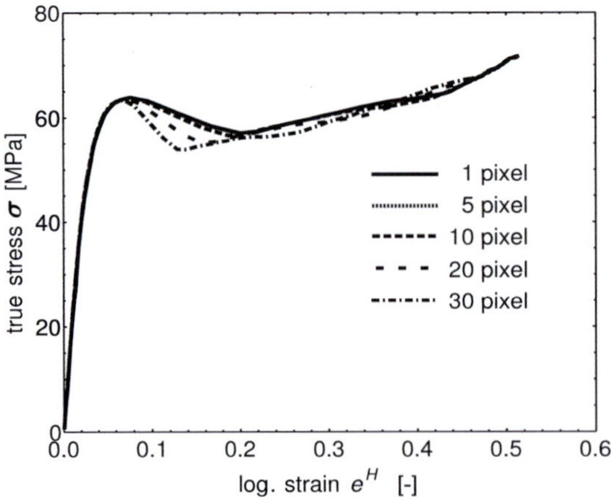

Figure 3.17.: Influence of the size of the "strain gauge" on the evaluation of uniaxial tensile tests of PC

As mentioned above, the strain provided by the correlation software Istra 4D is the Green-Lagrange strain tensor (2.16). However, in this work the logarithmic strain tensor $\boldsymbol{e}^H$ (2.21) is used. To obtain this tensor from the measured data, the Green-Lagrange strain tensor is reformulated in terms of the right stretch tensor $\boldsymbol{U}$

$$\boldsymbol{U} = \sqrt{2\boldsymbol{E} + \boldsymbol{1}} \tag{3.7}$$

and with (2.20) the logarithmic strain in the reference configuration is

$$\boldsymbol{E}^H = \ln \sqrt{2\boldsymbol{E} + \boldsymbol{1}} \quad . \tag{3.8}$$

In case of a uniaxial tensile test, the loading direction is a principal direction and no rotation occurs ($\boldsymbol{R} = \boldsymbol{1}$) so that with (2.12) it follows the identity of $\boldsymbol{e}^H = \boldsymbol{E}^H$ with $\boldsymbol{e}^H = \ln \boldsymbol{V}$ defined in (2.21). In a 2D DIC the components of $\boldsymbol{E}$ in loading direction (1-dir) and perpendicular to that direction (2-dir) are obtained only. The components of $\boldsymbol{e}^H$ in these directions are

$$e_1^H = \ln \sqrt{2E_1 + 1} \quad , \quad e_2^H = \ln \sqrt{2E_2 + 1} \quad . \tag{3.9}$$

The true stress in loading direction is computed from the force f_1 measured by the load cell and the current area of the cross section of the specimen as

$$\sigma_1 = \frac{f_1}{A} \quad . \tag{3.10}$$

In a 2D DIC the thickness t of the specimen is not measurable at the same time as the length l and width w; this would require a 3D DIC system with two cameras. Therefore, the assumption of the equal transverse strains $e_2^H = e_3^H$ which holds for isotropic materials is used with which the current cross section is then given by

$$A = w \cdot t = w_0 \cdot t_0 \cdot \exp[2e_2^H] \tag{3.11}$$

where $w = w_0 e_2^H$ and $t = t_0 e_3^H$.

Since the experiments are conducted at different strain rates it is necessary to determine the crosshead velocity v_{ch} for each desired strain rate before the tests. The strain rate is defined in the general form as

$$\dot{\varepsilon} = \frac{\mathrm{d}\varepsilon}{\mathrm{d}t} \tag{3.12}$$

where ε is a strain measure. With the nominal strain $\varepsilon^N = (l - l_0)/l_0$, the nominal strain rate $\dot{\varepsilon}^N$ reads as

$$\dot{\varepsilon}^N = \frac{\mathrm{d}(l - l_0)/l_0)}{\mathrm{d}t} = \frac{1}{l_0}\frac{\mathrm{d}l}{\mathrm{d}t} \tag{3.13}$$

where l_0 is the initial length and l the current length of the parallel region of the respective tensile test specimen. The crosshead velocity is then given by

$$v_{ch} = \frac{\mathrm{d}l}{\mathrm{d}t} = \dot{\varepsilon}^N l_0 \tag{3.14}$$

and is constant over the total experimental time. The nominal strain rate differs from the log. strain rate

$$\dot{\varepsilon}^H = \frac{\mathrm{d}e_1^H}{\mathrm{d}t} = \frac{\mathrm{d}\ln[l^*/l_0^*]}{\mathrm{d}t} = \frac{1}{l^*}\frac{\mathrm{d}l^*}{\mathrm{d}t} \tag{3.15}$$

due to the varying length l^* of the evaluation regions, for example these of Fig. 3.16. Depending on this length ($l^* \mathrel{\hat{=}}$ amount of pixel), the difference between the log. strain rates of the different strain gauges and the nominal strain rate is plotted in Fig. 3.18. While for constant velocity the nominal strain rate remains constant by virtue of (3.14), the log. strain rate is strongly dependent on the gauge length due to strain concentrations such as necking.

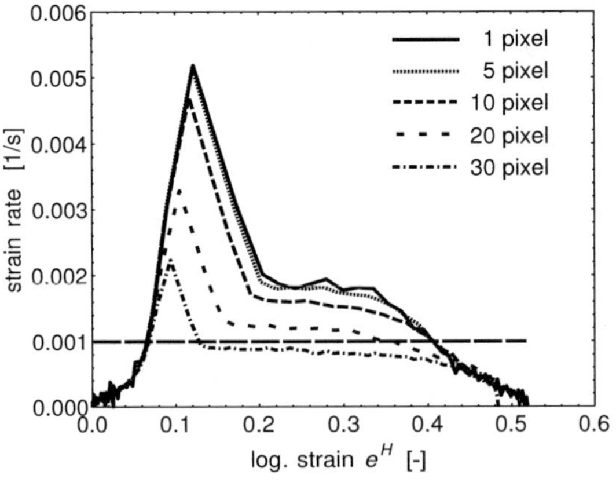

Figure 3.18.: The log. strain rate behavior of the PC test specimen subjected to different strain gauge sizes at a nominal strain rate of $\dot{\varepsilon}^N = 0.001\mathrm{s}^{-1}$ and a temperature of 20 °C

The true strain rates for PMMA at temperatures of 60 °C and 80 °C and a nominal strain rate of $\dot{\varepsilon}^N = 0.1\mathrm{s}^{-1}$ are shown in Fig. 3.19. It is observable that the true strain rate approaches the nominal strain rate with increasing temperature due to a more homogeneous deformation behavior over the total length of the test specimen at higher temperatures.

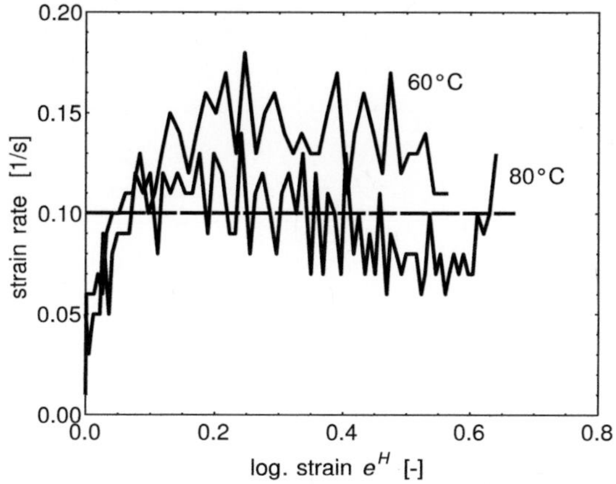

Figure 3.19.: The log. strain rate behavior of the PMMA test specimen at temperatures of 60 °C and 80 °C and a nominal strain rate of $\dot{\varepsilon}^N = 0.1\text{s}^{-1}$

Fig. 3.18 and Fig. 3.19 show a maximum deviation of a factor of five between the true strain rate and the nominal strain rate at a temperature of 20 °C. The difference between the nominal and the log. strain rate is thus neglected in the following experiments where the strain rates are varied over two orders of magnitude ($0.001 \ldots 0.1\text{s}^{-1}$).

3.2.3. Experimental findings for PMMA and PC

In the following experiments, three different constant nominal strain rates are applied on both materials: 0.1 s^{-1}, 0.01 s^{-1} and 0.001 s^{-1}. Since the materials differ in their glass transition temperature, different temperature ranges are considered in the experiments (Tab. 3.1). All experiments are repeated three times but only one representative curve is shown in the following since the amount of scatter was rather small. While both materials are investigated under monotonic loading, the behavior of PMMA is additionally studied under cyclic loading conditions.

Table 3.1.: Temperatures uniaxial tensile tests on PMMA and PC at different strain rates
(• ≙ experiment conducted, ∘ ≙ no experimental data obtained)

	PMMA ($T_g \approx 105$ °C)			PC ($T_g \approx 140$ °C)		
temperature [°C]	strain rate [s^{-1}]					
	$\dot{\varepsilon} = 0.001$	$\dot{\varepsilon} = 0.01$	$\dot{\varepsilon} = 0.1$	$\dot{\varepsilon} = 0.001$	$\dot{\varepsilon} = 0.01$	$\dot{\varepsilon} = 0.1$
20	•	•	•	•	•	•
60	•	•	•			
80	•	•	•	•	•	•
100	•	•	•			
105	∘	•	•			
110	∘	•	•			
120	∘	•	•	•	•	•
140				•	•	•
150				•	•	•

Monotonic tests

The true stress-log. strain curves for the three different strain rates are presented for PMMA in Fig. 3.20, and for PC in Fig. 3.21. All tests are performed until the test specimen fails or in case of higher temperatures the material yields in the clamped region. It is noted that at the lowest strain rate above 100 °C no valid experimental data is obtained for PMMA due to the very soft material response.

The temperature dependency is clearly observable for both materials and the glassy and rubbery state can be identified. Furthermore, not only the yield stress is temperature dependent but also the hardening modulus at higher strains decreases with increasing temperature. The failure strain is temperature dependent as well. The higher the temperature the higher the failure strain. It is noteworthy that in case of PMMA at a temperature of 20 °C brittle fracture in the elastic region occurs for the medium and high strain rate while at the lower strain rate the material behaves ductile. In contrast, PC behaves ductile at room temperature even for the highest strain rate considered (Senden et al., 2012).

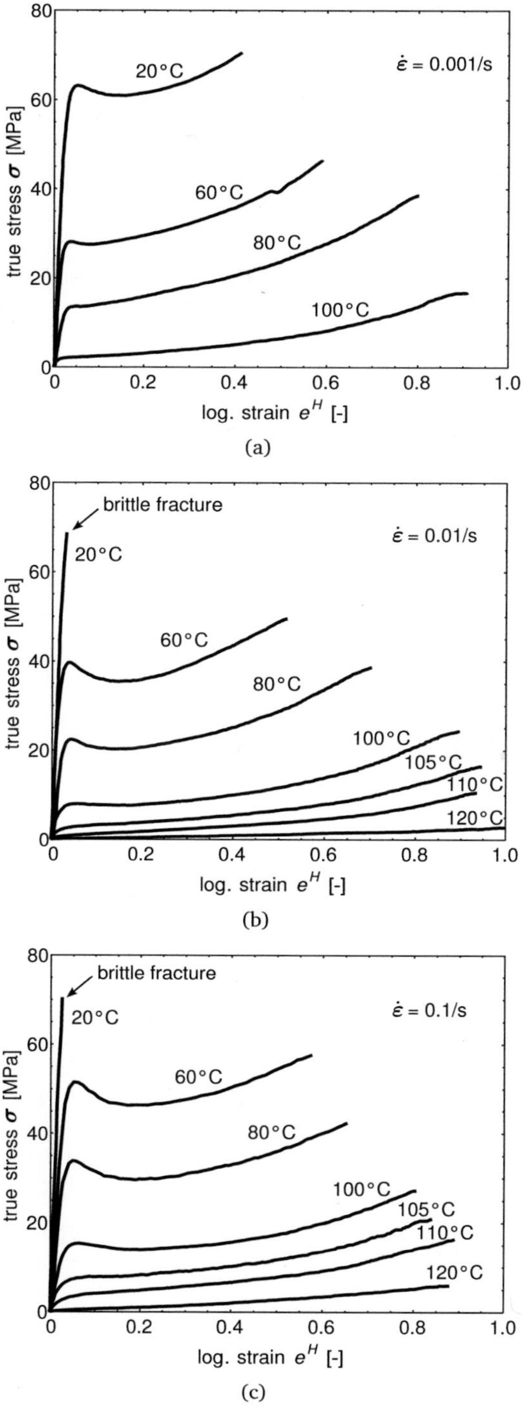

Figure 3.20.: Uniaxial tensile true stress-log. strain curves of PMMA at three different strain rates and various temperatures

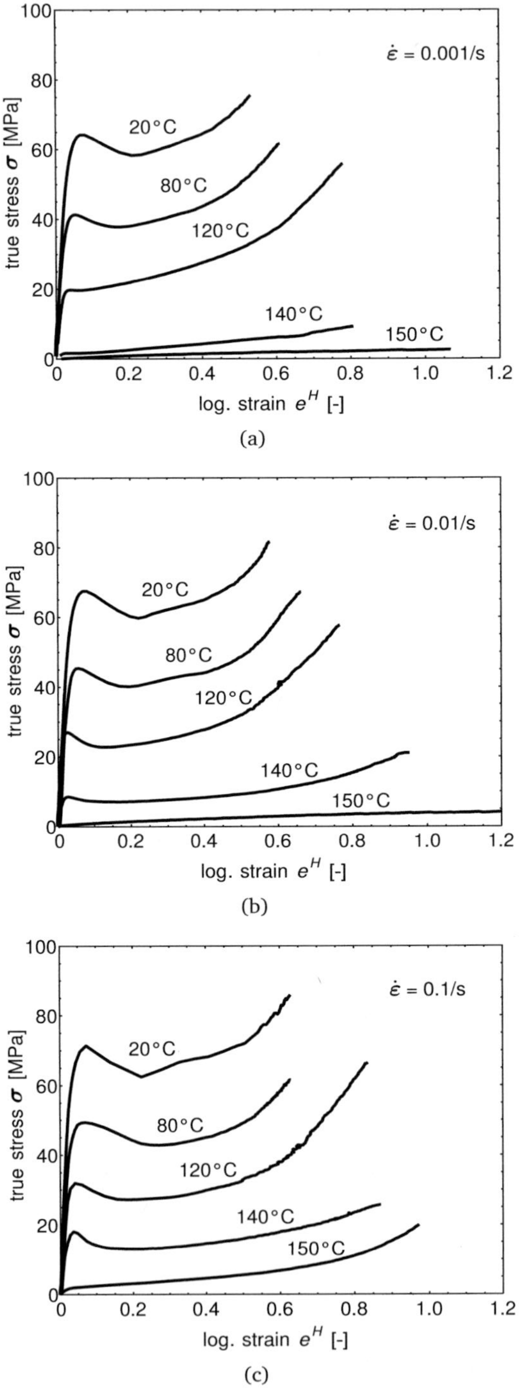

Figure 3.21.: Uniaxial tensile true stress-log. strain curves of PC at three different strain rates and various temperatures

In Fig. 3.22, the comparison of the true stress-log. strain curves at high and low strain rates for both materials is shown. Of course, the materials exhibit a strong strain rate dependency which is much more pronounced for PMMA than for PC. The strain rate affects not only the onset of yielding but also the hardening of the material at higher strains which can be seen clearly at temperatures below the glass transition. At higher strain rates adiabatic heating of the specimen occurs that reduces the hardening slope[6] (Arruda et al., 1995). At lower strain rates less heating emerges and may be conducted due to a longer time period. This more isothermal deformation process yields a more pronounced hardening and in case of PC the stress-strain curves are even crossing each other. Furthermore, it can be observed by means of the yield stress that at a temperature of 100 °C PMMA at the low strain rate is in the rubbery state already while at the higher strain rate it is still in the glassy state. This clearly indicates that the glass transition temperature is rate dependent as well which is well-known from dynamical mechanical thermal analyses (DMTA) experiments.

The strain rate dependency of the hardening behavior is most significant at temperatures around the glass transition as shown in Fig. 3.23. It can be observed that the hardening modulus decreases with increasing temperature and increasing strain rate.

A more detailed dependence of PMMA on the strain rate and temperature in the small strain region is depicted in Fig. 3.24[7]. The amount of softening (Fig. 3.24c) is defined as the ratio σ_0/σ_p of the yield peak stress σ_0 to the stress minimum (plateau) σ_p before hardening occurs. Young's modulus shows the viscoelasticity of the material as well as the typical drop in the region of the glass transition (cf. Fig. 3.3). The yield stress, taken here as the minimum stress after softening, decreases approximately linearly with increasing temperature till the glass transition is reached and the material enters the rubbery state. In Fig. 3.24c, the strain rate and temperature dependency of the softening is clearly observable. It tends to vanish for all strain rates when the temperature reaches the glass transition temperature.

[6] The effect of a reduced hardening modulus at higher strain rates complicates the usage of these data for modeling and fitting unless one models this effect as well by computing the temperature increase by virtue the energy balance and consider the plastic work generating heat (Arruda et al., 1995). For this reason, even lower strain rates may be preferred, but this would increase the time for the experiments enormously.

[7] Note, the measurement of the DIC is rather inaccurate in the small strain response.

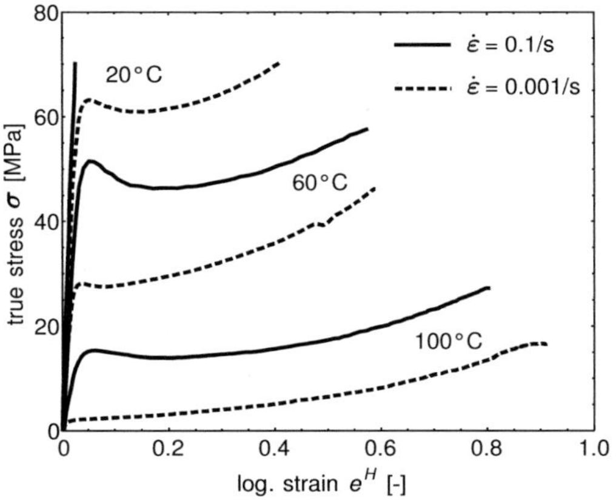

(a) Strain rate and temperature dependency of PMMA

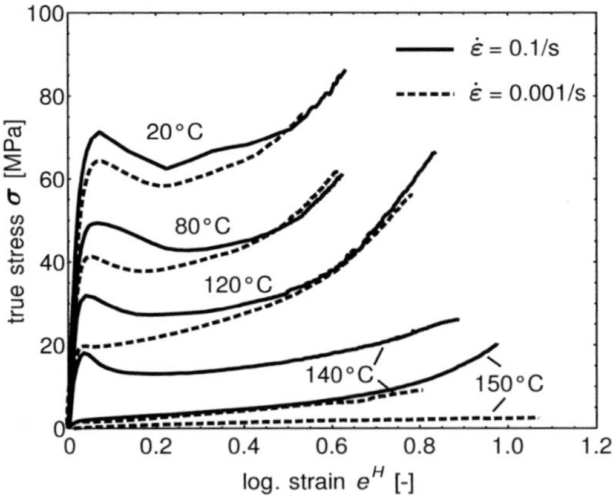

(b) Strain rate and temperature dependency of PC

Figure 3.22.: Tensile true stress - log. strain curves at higher and lower strain rate at different temperatures

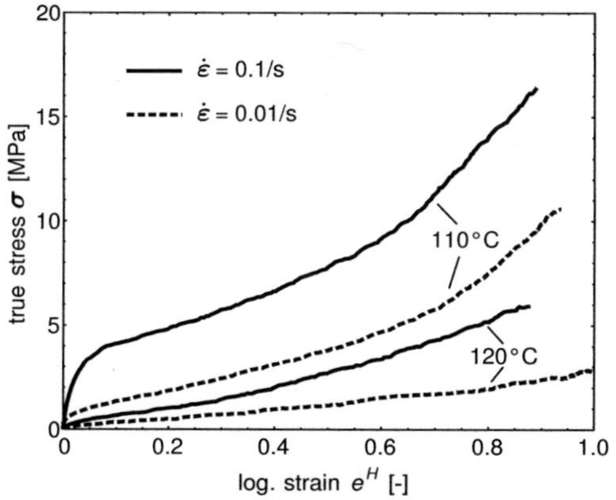

(a) PMMA above glass transition ($T_g = 105$ °C)

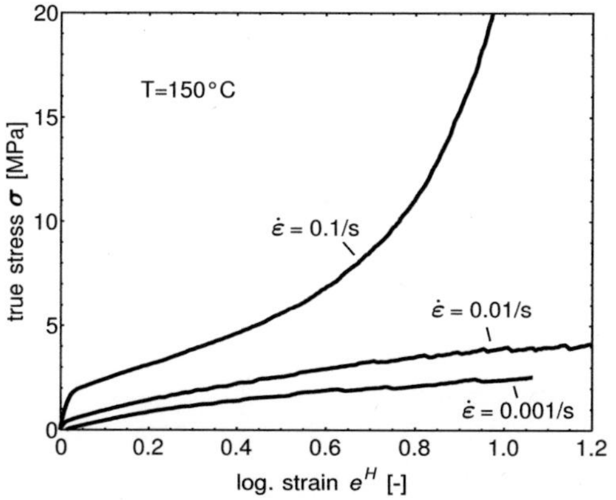

(b) PC above glass transition ($T_g = 145$ °C)

Figure 3.23.: Tensile true stress-log. strain curves at temperatures above glass transition

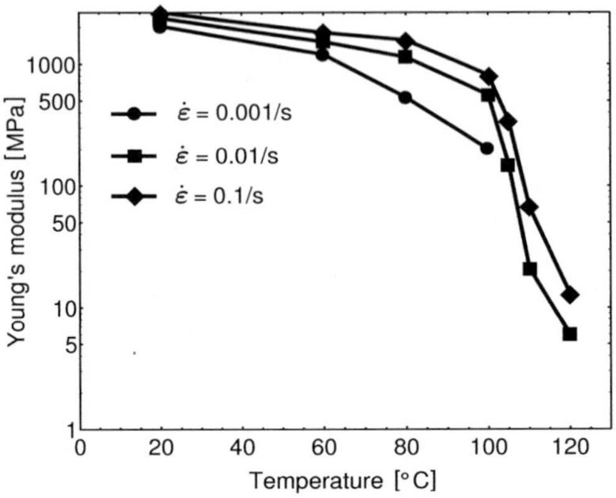

(a) Young's modulus as a function of temperature

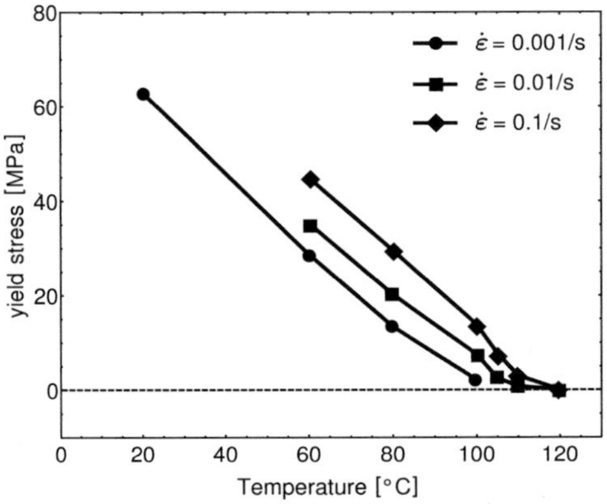

(b) Yield stress as a function of temperature

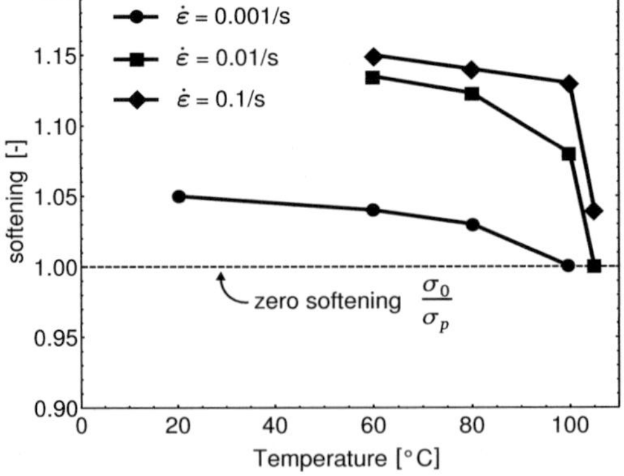

(c) Softening behavior as a function of temperature

Figure 3.24.: Strain rate and temperature dependency of PMMA in small strain region

Cyclic tests on PMMA

The cyclic uniaxial tensile tests on PMMA are conducted by applying the same histories of overall specimen displacement for all temperatures and strain rates. This leads to different strain values where the unloading takes place because the deformation is more homogeneous at elevated temperatures and thus the local strain concentration is smaller. The unloading displacement is driven to the point where the force is equal to zero. Hence, no compressive stress is measured. In Tab. 3.2 the series of experimental tests is summarized.

Table 3.2.: Temperatures and strain rates of the cyclic uniaxial tensile tests on PMMA (• $\hat{=}$ experiment conducted, ∘ $\hat{=}$ no experimental data obtained)

temperature [°C]	strain rate [s^{-1}]		
	$\dot{\varepsilon} = 0.001$	$\dot{\varepsilon} = 0.01$	$\dot{\varepsilon} = 0.1$
60	•	•	•
80	•	•	•
100	•	•	•
105	∘	•	•
110	∘	∘	•

In Fig. 3.25 the stress-strain curves of the cyclic tensile tests on PMMA are shown for three different strain rates and various temperatures. For increasing strain the material behaves like in the monotonic tensile test but during unloading a clear hysteresis is observable. The area of the hysteresis increases with increasing temperature and decreases with increasing strain rate due to the viscous response of the material. Furthermore, the hysteresis area increases at larger strains and the unloading-reloading path is much more non-linear than for smaller strains. This effect is hardly reported in the literature and is not understood yet.

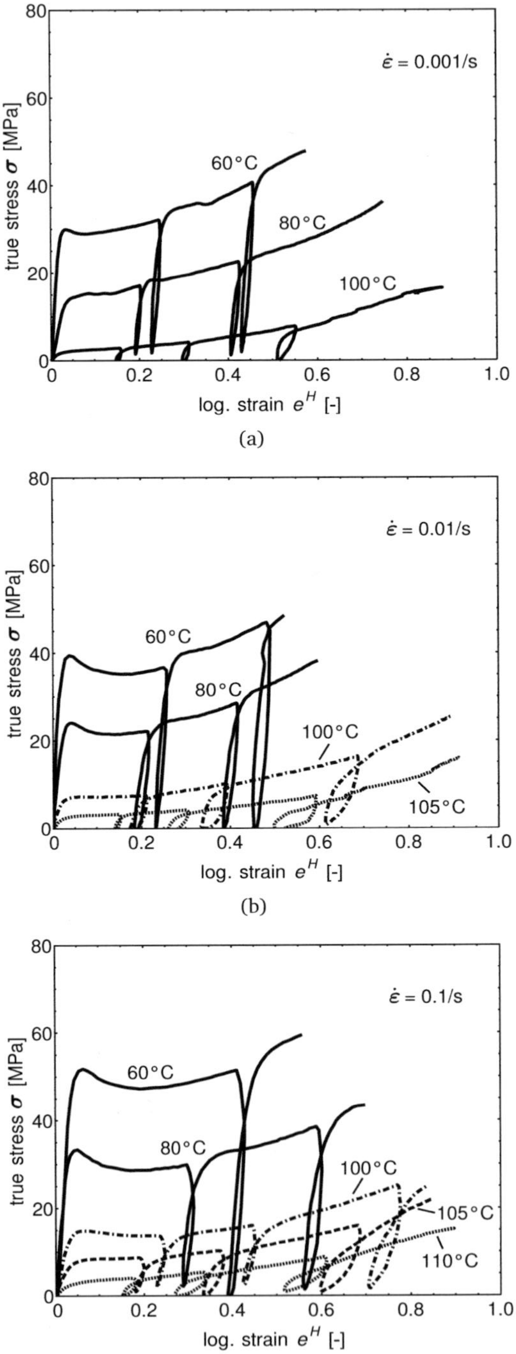

(a)

(b)

(c)

Figure 3.25.: Cyclic true stress-log. strain curves of PMMA at three different strain rates and various temperatures

4. Modeling the mechanical behavior of amorphous thermoplastic polymers

In this chapter constitutive models for the class of amorphous thermoplastic polymers will be established in order to capture the mechanical behavior characterized in the previous chapter. In a first step, it will be shown that the classical model for glassy polymers by Boyce et al. (1988) is insufficient for temperatures near or and above the glass transition. Based on these results, the model will be enhanced in several regards to be valid at elevated temperatures as well. This is done aiming at a minimum number of material parameters necessary to describe the thermomechanical behavior sufficiently. The ability of two different models accounting for the temperature-dependent molecular network, and thus the hardening behavior at large strains, will be investigated. The models are fitted to the experimental data of PMMA given in the previous chapter and are analyzed by subjecting them to a fictitious thermomechanical loading history. Additionally, the models will be further modified by incorporating an initial plastic stretch tensor representing a processing induced microstructure in terms of a "frozen-in" pre-stretch of the molecular network.

4.1. Three dimensional finite strain model for glassy polymers

The experimental findings of the previous chapter determine the features of the model required to reproduce the mechanical behavior. In detail, the model shall display isotropic elasticity, temperature and strain rate-dependent yield and hardening at large strains. In the domain of large strains the small strain viscoelasticity is not important and can be neglected. The intrinsic softening at the onset of yield depends on thermal loading history and is not present at elevated temperatures anymore (cf. Ch. 3). Thus, it is not that important in simulations where a thermomechanical load is applied. Due to this reason modeling of the intrinsic softening is not considered in this work[1]. The strain rate dependence of the glass transition is not considered as well since it is expected to be insignificant in the regarded strain rate region.

The well established and widely used constitutive model for glassy polymers proposed by Boyce et al. (1988) is adopted to provide a basic model. Yet, it does not reproduce all of the desired properties as shown in this section. The model accounts for two resistances against plastic deformation (Sec. 3.1.2); the intermolecular resistance against molecular chain segment rotations and the *entropic* intramolecular resistance against molecular chain alignment. Originally, both resistances have been modeled by Haward and Thackray (1968) in 1D. They utilized an Eyring dashpot to describe the initiation of flow and a highly non-linear Langevin spring for the resistance against the stretching of the chains derived from the non-Gaussian statistical mechanics of rubber elasticity (Treloar, 1974). In order to improve the description of the rate dependency of yielding the micromechanically motivated "double-kink model" has been developed by Argon (1973). This double-kink model is used by Boyce et al. (1988) in combination with the rubber elastic three-chain model (Wang and Guth, 1952) for the entropic network deformation resistance in a 3D continuum mechanical framework. The overall model includes the effects of rate-, temperature- and pressure-dependent plastic flow, softening and a fairly realistic hardening behavior at large strains well below the glass transition temperature. Arruda and Boyce (1993) developed the rubber elastic eight-chain model and incorporated it in their model for glassy polymers to capture the hardening behavior more realistically (Arruda et al., 1993).

The rheological model in Fig. 4.1 illustrates the components of the basic model of this work. It consists of an elastic and a visco-plastic part. The elastic behavior of the material is modeled by a non-linear (hyperelastic) spring. The visco-plastic part is given by a dashpot representing the rate- and temperature-dependent intermolecular resistance against plastic flow and a second, likewise non-linear spring for the resistance against the molecular network deformation. As a result of the dashpot, the model is a non-linear viscoelastic model but describes plastic deformations due to a strong non-linear formulation. The decomposition in an elastic and a plastic part is only admissible for temperatures far below the glass transition, since the *permanent* stretch and orientation of the molecular network is reversible if the polymer is heated up above the glass transition. However, the split into an elastic and an inelastic part remains valid for all temperatures.

[1] Approaches to model the intrinsic softening can be found for example in Boyce et al. (1988), Hasan et al. (1993) or Klompen et al. (2005).

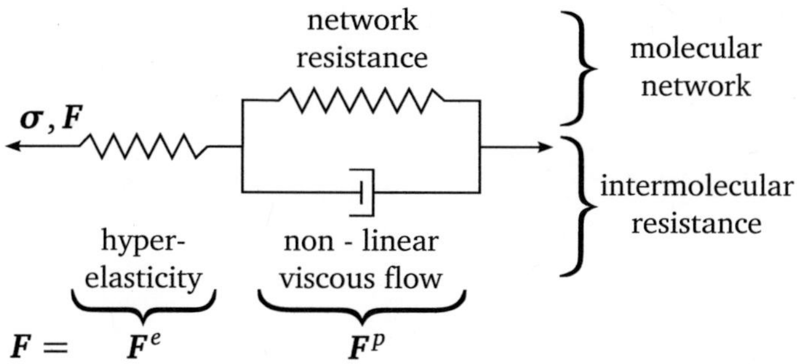

Figure 4.1.: Rheological model

The constitutive equations are formulated in a three dimensional non-linear continuum mechanical framework obeying the standard requirements of requirements of frame-indifference and thermodynamic consistency (Holzapfel, 2007). The kinematics and the constitutive equations are presented in detail in the following sections.

4.1.1. Kinematics

As shown in Sec. 2.1, the kinematics is generally described by the deformation gradient $\boldsymbol{F}$ (2.4). Now, it is necessary to distinguish between elastic and inelastic deformation. For this purpose, it is assumed that the deformation gradient may be decomposed multiplicatively as (Lee, 1969)

$$\boldsymbol{F} = \boldsymbol{F}^e \boldsymbol{F}^p \tag{4.1}$$

with the elastic $\boldsymbol{F}^e$ and the inelastic $\boldsymbol{F}^p$ parts. This implies an existence of a local stress free intermediate configuration $\hat{\mathcal{B}}$ obtained after elastically unloading by $\boldsymbol{F}^{e-1}$. In the context of polymers, $\boldsymbol{F}^p$ physically represents the *permanent* molecular orientation and stretch in the material. This is schematically shown in Fig. 4.2 where the state of the molecular network in each configuration is sketched. While the network is initially isotropic in the reference configuration, it is oriented and stretched in the loaded current configuration as well as in the intermediate configuration. The latter two configurations differ only in the amount of the elastic deformation.

With the above decomposition of the deformation gradient, the spatial velocity gradient (2.41) is written as

$$\boldsymbol{l} = \dot{\boldsymbol{F}}\boldsymbol{F}^{-1} = \dot{\boldsymbol{F}}^e\boldsymbol{F}^{e-1} + \boldsymbol{F}^e\dot{\boldsymbol{F}}^p\boldsymbol{F}^{p-1}\boldsymbol{F}^{e-1} = \dot{\boldsymbol{F}}^e\boldsymbol{F}^{e-1} + \boldsymbol{F}^e\hat{\boldsymbol{l}}^p\boldsymbol{F}^{e-1} = \boldsymbol{l}^e + \boldsymbol{l}^p \tag{4.2}$$

where $\hat{\boldsymbol{l}}^p = \dot{\boldsymbol{F}}^p\boldsymbol{F}^{p-1}$ is the inelastic velocity gradient in the intermediate configuration and $\boldsymbol{l}^p = \boldsymbol{F}^e\hat{\boldsymbol{l}}^p\boldsymbol{F}^{e-1}$ that in the current configuration, respectively. Analogous to the

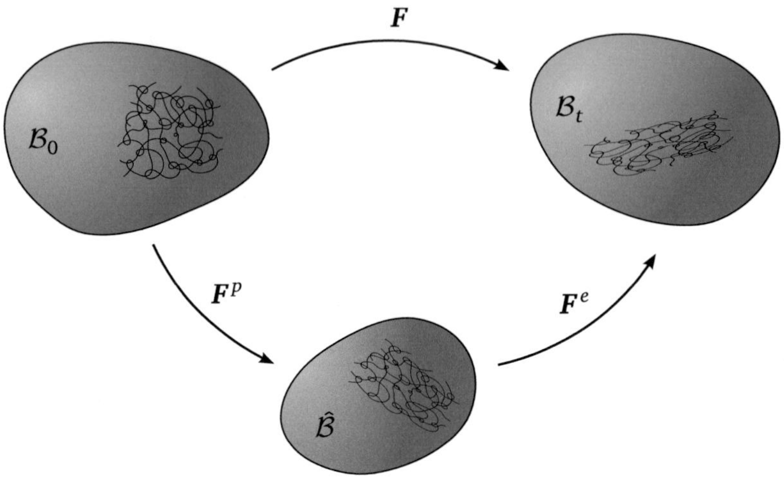

Figure 4.2.: Split of the deformation gradient and the resulting intermediate configuration $\hat{\mathcal{B}}$

additive split of the velocity gradient (2.41) into a symmetric and skew-symmetric part, the inelastic velocity gradient is split into

$$\hat{\boldsymbol{l}}^p = \hat{\boldsymbol{d}}^p + \hat{\boldsymbol{w}}^p \tag{4.3}$$

with the symmetric inelastic rate of deformation tensor $\hat{\boldsymbol{d}}^p$ and the skew-symmetric inelastic spin tensor $\hat{\boldsymbol{w}}^p$, both belong to the intermediate configuration. While $\hat{\boldsymbol{d}}^p$ has to be constitutively prescribed by a flow rule, the inelastic spin can be either prescribed or be computed. The latter can be done by imposing a restriction on the undetermined elastic and inelastic rotation tensors in the polar decomposition (Boyce et al., 1989b)

$$\boldsymbol{F} = \boldsymbol{F}^e \boldsymbol{F}^p = \boldsymbol{V}^e \boldsymbol{R}^e \boldsymbol{R}^p \boldsymbol{U}^p = \boldsymbol{V}^e \boldsymbol{R} \boldsymbol{U}^p \quad . \tag{4.4}$$

Assigning all rotations to the intermediate configuration with $\boldsymbol{R} = \boldsymbol{R}^p$ and $\boldsymbol{R}^e = \boldsymbol{1}$ results in a symmetric elastic deformation gradient

$$\boldsymbol{F}^e = \boldsymbol{R}^e \boldsymbol{U}^e = \boldsymbol{U}^e = \boldsymbol{V}^e \boldsymbol{R}^e = \boldsymbol{V}^e = \boldsymbol{F}^{eT} \tag{4.5}$$

$$\boldsymbol{F}^p = \boldsymbol{R}^p \boldsymbol{U}^p = \boldsymbol{V}^p \boldsymbol{R}^p = \boldsymbol{R} \boldsymbol{U}^p = \boldsymbol{V}^p \boldsymbol{R} \quad . \tag{4.6}$$

In order to algebraically prescribe $\hat{\boldsymbol{w}}^p$, the result of (4.5) is utilized, i.e. an equation in terms of $\boldsymbol{F}^e$ must be found first (Boyce et al., 1989b). Using (2.38), (4.2) and (4.3) and rearranging in terms of $\dot{\boldsymbol{F}}^e$ and $\dot{\boldsymbol{F}}^{eT}$ we get

$$\dot{\boldsymbol{F}}^e = (\boldsymbol{d} + \boldsymbol{w}) \boldsymbol{F}^e - \boldsymbol{F}^e (\hat{\boldsymbol{d}}^p + \hat{\boldsymbol{w}}^p) \quad , \quad \dot{\boldsymbol{F}}^{eT} = \boldsymbol{F}^{eT} (\boldsymbol{d} - \boldsymbol{w}) - (\hat{\boldsymbol{d}}^p - \hat{\boldsymbol{w}}^p) \boldsymbol{F}^{eT} \quad . \tag{4.7}$$

With (4.5) it is also $\dot{\boldsymbol{F}}^{e} = \dot{\boldsymbol{F}}^{eT}$, which gives

$$(\boldsymbol{d}+\boldsymbol{w})\boldsymbol{F}^e - \boldsymbol{F}^e(\hat{\boldsymbol{d}}^p + \hat{\boldsymbol{w}}^p) = \boldsymbol{F}^e(\boldsymbol{d}-\boldsymbol{w}) - (\hat{\boldsymbol{d}}^p - \hat{\boldsymbol{w}}^p)\boldsymbol{F}^e \tag{4.8}$$

and with resorting

$$(\hat{\boldsymbol{d}}^p + \boldsymbol{d})\boldsymbol{F}^e - \boldsymbol{F}^e(\hat{\boldsymbol{d}}^p + \boldsymbol{d}) = \boldsymbol{F}^e(\hat{\boldsymbol{w}}^p - \boldsymbol{w}) + (\hat{\boldsymbol{w}}^p - \boldsymbol{w})\boldsymbol{F}^e \quad . \tag{4.9}$$

The solution of the above algebraic equation for $\hat{\boldsymbol{w}}^p$ in terms of the other kinematic quantities is given in Agah-Tehrani et al. (1987) as

$$\hat{\boldsymbol{w}}^p = \boldsymbol{w} + \mathrm{tr}[\boldsymbol{z}]\,\boldsymbol{g} - (\boldsymbol{g}\boldsymbol{z} + \boldsymbol{z}\boldsymbol{g}) \tag{4.10}$$

$$\text{with } \boldsymbol{g} = (\hat{\boldsymbol{d}}^p + \boldsymbol{d})\boldsymbol{F}^e - \boldsymbol{F}^e(\hat{\boldsymbol{d}}^p + \boldsymbol{d}) \tag{4.11}$$

$$\text{and } \boldsymbol{z} = (\mathrm{tr}[\boldsymbol{F}^e]\boldsymbol{1} - \boldsymbol{F}^e)^{-1} \quad . \tag{4.12}$$

As mentioned previously, the second possibility to determine the inelastic spin is to prescribe it. The simplest choice is a non spinning intermediate configuration, i.e.

$$\hat{\boldsymbol{w}}^p = \boldsymbol{0} \quad . \tag{4.13}$$

Both representations of the inelastic spin, (4.10) and (4.13), result in an equivalent model response[2] (Boyce et al., 1989b). However, the numerical treatment of the model is different, since the update of the symmetric elastic deformation gradient yields only six equations. However, the constraint $\boldsymbol{R}^e = \boldsymbol{1}$ then is not enforced exactly by the integration algorithm so that further equations must be introduced to solve this problem accurately (Holopainen and Wallin, 2013). To overcome this drawback and for a simpler set of equations, the choice (4.13) is made for the inelastic spin in the following.

4.1.2. Constitutive equations

In this section the overall structure of the constitutive model consistent with the second law of thermodynamics is worked out first. This implies for thermodynamic consistency that the internal dissipation (2.92) must be non-negative at all times. For this purpose, the Helmholtz free energy Ψ per unit mass, i.e. the stored energy in the material, is additively split into an elastic Ψ^e and an inelastic part Ψ^p (e.g. Kamlah (1994), Holzapfel (2007),

[2] This was shown by comparing the stress-strain response of the model using the two different representations of the inelastic spin in a computational example (Boyce et al., 1989b).

Neto et al. (2008)). Besides the total deformation $\boldsymbol{F}$, it is assumed that the state of the material and thus the elastic stored energy depends on the elastic right $\hat{\boldsymbol{C}}^e = \boldsymbol{F}^{eT}\boldsymbol{F}^e$, the inelastic left Cauchy-Green tensor $\hat{\boldsymbol{b}}^p = \boldsymbol{F}^p\boldsymbol{F}^{pT}$ in the intermediate configuration and the temperature T. The free energy per unit mass with respect to tensors in the intermediate configuration than writes

$$\Psi = \Psi^{\dagger}(\hat{\boldsymbol{C}}^e, \hat{\boldsymbol{b}}^p, T) = \Psi^e(\hat{\boldsymbol{C}}^e, T) + \Psi^p(\hat{\boldsymbol{b}}^p, T) \tag{4.14}$$

and comprises the stored elastic energy Ψ^e and the inelastic energy Ψ^p due to stretching the molecular chains. Performing the time derivative of the free energy

$$\dot{\Psi} = \frac{\partial \Psi^e}{\partial \hat{\boldsymbol{C}}^e} : \dot{\hat{\boldsymbol{C}}}^e + \frac{\partial \Psi^e}{\partial T}\dot{T} + \frac{\partial \Psi^p}{\partial \hat{\boldsymbol{b}}^p} : \dot{\hat{\boldsymbol{b}}}^p + \frac{\partial \Psi^p}{\partial T}\dot{T} \tag{4.15}$$

by using the relations $\dot{\hat{\boldsymbol{C}}}^e = 2\boldsymbol{F}^{eT}\boldsymbol{d}^e\boldsymbol{F}^e$ and $\dot{\hat{\boldsymbol{b}}}^p = \hat{\boldsymbol{l}}^p\hat{\boldsymbol{b}}^p + \hat{\boldsymbol{b}}^p\hat{\boldsymbol{l}}^{pT}$ and additionally exploiting the tensor algebra rules for symmetric tensors [3] we get

$$\dot{\Psi} = 2\boldsymbol{F}^e\frac{\partial \Psi^e}{\partial \hat{\boldsymbol{C}}^e}\boldsymbol{F}^{eT} : \boldsymbol{d}^e + 2\hat{\boldsymbol{b}}^p\frac{\partial \Psi^p}{\partial \hat{\boldsymbol{b}}^p} : \hat{\boldsymbol{d}}^p + (\frac{\partial \Psi^e}{\partial T} + \frac{\partial \Psi^p}{\partial T})\dot{T} \quad . \tag{4.16}$$

Inserting this result in (2.93) and making use of $\boldsymbol{d} = \boldsymbol{d}^e + \boldsymbol{d}^p$ with $\boldsymbol{d}^p = sym[\boldsymbol{F}^e\hat{\boldsymbol{l}}^p\boldsymbol{F}^{e-1}]$ leads to

$$\begin{aligned}(\boldsymbol{\tau} - 2\rho_0\boldsymbol{F}^e\frac{\partial \Psi^e}{\partial \hat{\boldsymbol{C}}^e}\boldsymbol{F}^{eT}) : \boldsymbol{d} + (2\rho_0\boldsymbol{F}^{eT}\boldsymbol{F}^e\frac{\partial \Psi^e}{\partial \hat{\boldsymbol{C}}^e}\boldsymbol{F}^{eT}\boldsymbol{F}^{e-T} - 2\rho_0\hat{\boldsymbol{b}}^p\frac{\partial \Psi^p}{\partial \hat{\boldsymbol{b}}^p}) : \hat{\boldsymbol{d}}^p \\ -(\frac{\partial \Psi^e}{\partial T} + \frac{\partial \Psi^p}{\partial T} + s)\rho_0\dot{T} \geq 0 \quad .\end{aligned} \tag{4.17}$$

The above equation must be true for all feasible thermomechanical processes, for example a pure elastic process ($\hat{\boldsymbol{d}}^p = \boldsymbol{0}$) at which the internal dissipation vanishes identically. This implies potential relations for the entropy $s = -(\frac{\partial \Psi^e}{\partial T} + \frac{\partial \Psi^p}{\partial T})$ and the Kirchhoff stress (Coleman and Gurtin, 1967)

$$\boldsymbol{\tau} = 2\rho_0\boldsymbol{F}^e\frac{\partial \Psi^e}{\partial \hat{\boldsymbol{C}}^e}\boldsymbol{F}^{eT} = \boldsymbol{F}^e\hat{\boldsymbol{S}}\boldsymbol{F}^{eT} \tag{4.18}$$

[3] The free energy parts Ψ^e and Ψ^p are assumed to be isotropic functions of $\hat{\boldsymbol{C}}^e$ and $\hat{\boldsymbol{b}}^p$, respectively. A consequence of this assumption is the coaxiality of $\frac{\partial \Psi^e}{\partial \hat{\boldsymbol{C}}^e}$ and $\hat{\boldsymbol{C}}^e$ and of $\frac{\partial \Psi^p}{\partial \hat{\boldsymbol{b}}^p}$ and $\hat{\boldsymbol{b}}^p$. Performing the tensor products in (4.16) and (4.17) results in symmetric tensors. In this special case, the Mandel stress tensor is symmetric as well which must not be true in general.

with the second Piola-Kirchhoff stress tensor in the intermediate configuration $\hat{\boldsymbol{S}}$. By defining the Mandel stress tensor

$$\boldsymbol{\Sigma} = \hat{\boldsymbol{C}}^e \hat{\boldsymbol{S}} = \boldsymbol{F}^{eT} \boldsymbol{\tau} \boldsymbol{F}^{e-T} \tag{4.19}$$

and the backstress tensor

$$\hat{\boldsymbol{\tau}}^b = 2\rho_0 \hat{\boldsymbol{b}}^p \frac{\partial \Psi^p}{\partial \hat{\boldsymbol{b}}^p} \tag{4.20}$$

we get the driving stress tensor

$$\boldsymbol{\Sigma}^* = \boldsymbol{\Sigma} - \hat{\boldsymbol{\tau}}^b \tag{4.21}$$

With these quantities, the remaining inequality (4.17), the internal dissipation, may be rewritten in a reduced form as

$$\mathcal{D}_{int} = \boldsymbol{\Sigma}^* : \hat{\boldsymbol{d}}^p \geq 0 \tag{4.22}$$

which provides a restriction for the constitutive equations.

In the following the elastic and inelastic parts of the free energy as well as a flow rule for the inelastic rate of deformation tensor consistent with (4.22) is specified to model both, the intermolecular and the network resistance.

Intermolecular resistance

The origin of the elastic stiffness and the yield strength of the polymer is caused by the intermolecular resistance as discussed in Sec. 3.1.2. The elasticity is modeled with a compressible Neo-Hooke model with a strain energy function $W^e = \rho_0 \Psi^e$ that depends on the invariants of the elastic right Cauchy-Green tensor in the intermediate configuration

$$W^e = \frac{\mu}{2}\left(\mathrm{I}_{\hat{C}^e} - 3\right) - \frac{\mu}{2}\ln\left(\mathrm{III}_{\hat{C}^e}\right) + \frac{\lambda}{2}\left(\sqrt{\mathrm{III}_{\hat{C}^e}} - 1\right)^2 \tag{4.23}$$

where μ and λ are the Lamé parameters[4]. While Boyce et al. (1988) focused only on glassy polymers and thus considered small elastic strains, in this work a more general approach is followed to capture large elastic strains at temperatures beyond the glass transition temperature.

[4] The Lamé parameters are given in terms of Young's modulus E and Poisson's ratio ν as $\lambda = \frac{\nu E}{(1+\nu)(1-2\nu)}$ and $\mu = \frac{E}{2(1+\nu)}$.

According to (4.18) the second Piola-Kirchhoff stress tensor with respect to the intermediate configuration is given by

$$\hat{\boldsymbol{S}} = 2\frac{\partial W^e}{\partial \hat{\boldsymbol{C}}^e} = \mu(\mathbf{1} - \hat{\boldsymbol{C}}^{e-T}) + \lambda(J^e - 1)\, J^e \hat{\boldsymbol{C}}^{e-T} \tag{4.24}$$

so that the Mandel stress tensor - likewise in the intermediate configuration - can be written as

$$\boldsymbol{\Sigma} = \hat{\boldsymbol{C}}^e \hat{\boldsymbol{S}} = \mu(\hat{\boldsymbol{C}}^e - \mathbf{1}) + \lambda(J^e - 1)\, J^e \mathbf{1} \tag{4.25}$$

and the Kirchhoff stress tensor in the current configuration as

$$\boldsymbol{\tau} = \mu(\boldsymbol{b}^e - \mathbf{1}) + \lambda(J^e - 1)\, J^e \mathbf{1} \quad . \tag{4.26}$$

Inelastic deformations are described through a flow rule for the inelastic rate of deformation tensor $\hat{\boldsymbol{d}}^p$ with the deviatoric driving stress $\boldsymbol{\Sigma}^{*\prime}$ and the inelastic shear strain rate $\dot{\gamma}^p$ as

$$\hat{\boldsymbol{d}}^p = \dot{\gamma}^p \frac{\boldsymbol{\Sigma}^{*\prime}}{||\boldsymbol{\Sigma}^{*\prime}||} \tag{4.27}$$

where $||\boldsymbol{\Sigma}^{*\prime}|| = (\mathrm{tr}[\boldsymbol{\Sigma}^{*\prime}\boldsymbol{\Sigma}^{*\prime}])^{1/2}$. Due to the deviatoric stress, this type of flow rule provides the inelastic incompressibility as experimentally observed for shear yielding in amorphous polymers. Inserting that flow rule in the dissipation inequality (4.22)

$$\mathcal{D}_{int} = \boldsymbol{\Sigma}^* : \dot{\gamma}^p \frac{\boldsymbol{\Sigma}^{*\prime}}{||\boldsymbol{\Sigma}^{*\prime}||} = \dot{\gamma}^p ||\boldsymbol{\Sigma}^{*\prime}|| \geq 0 \tag{4.28}$$

yields the restriction for the inelastic shear strain rate $\dot{\gamma}^p \geq 0$ for thermodynamic consistency. Note that this model does not contain a flow function as conventionally used for e.g. metal plasticity.

As discussed in Sec. 3.1.2 the viscoinelastic flow of amorphous glassy polymers is a thermally activated process. This is typically described by an Arrhenius-type equation for $\dot{\gamma}^p$ as

$$\dot{\gamma}^p = \dot{\gamma}_0^p \exp[\frac{-\Delta G}{k\, T}] \tag{4.29}$$

where $\dot{\gamma}_0^p$ is a reference inelastic strain rate, ΔG the stress-dependent activation energy, k the Boltzmann constant [5] and T the absolute temperature (Haward and Young, 1997). According to the "double-kink model" by Argon (1973) the activation energy is given by

$$\Delta G = \frac{3\pi\mu\omega^2 a^3}{16(1-\nu)} \left(1 - \left(\frac{||\boldsymbol{\Sigma}^{*\prime}||}{s_0}\right)^{5/6}\right) \tag{4.30}$$

[5] The value of the Boltzmann constant is $k = 1.380648 \cdot 10^{-23}$J/K

and describes the required energy for chain segment rotations against the elastic stiffness of surrounding chains. In (4.30) ω is the angle of rotation, a the mean radius of a molecular chain and $s_0 = 0.077\mu/(1-\nu)$ with the shear modulus μ and Poisson's ratio ν. The pre-factor is summarized to be one material parameter as $A = 39\pi\omega^2 a^3/16k$. In this work the exponent of 5/6 in (4.30) is neglected in view of its small influence and the resulting simplification of the equation. Furthermore, the so-called "athermal shear strength" s_0 is taken to be an independent material constant[6] and not to be dependent on the shear modulus μ as originally suggested by Argon (1973).

Due to the exponential function in (4.29), the condition of $\dot{\gamma}^p(\Sigma = 0) = 0$ is not fulfilled exactly when (4.30) in combination with (4.29) is used with no further modifications. Thus, the function of (4.29) with (4.30) is subtracted for a zero stress state ($\|\Sigma^{*\prime}\| = 0$). This leads to the following expression for the inelastic shear strain rate

$$\dot{\gamma}^p = \dot{\gamma}_0^p \left(\exp\left[\frac{A}{T}(\|\Sigma^{*\prime}\| - s) \right] - \exp\left[-\frac{A}{T} s \right] \right) \tag{4.31}$$

which fulfills the restriction of (4.28) if $\dot{\gamma}_0^p \geq 0$ holds. In the above equation the pressure dependence of the yield stress typically observed in polymers, e.g. Kinloch and Young (1983), is incorporated by $s = s_0 + \alpha p$ where α is a material parameter controlling the tensile-compression asymmetry and $p = -\frac{1}{3}\mathrm{tr}[\Sigma]$ is the hydrostatic pressure.

Molecular network resistance

The inelastic part of the free energy Ψ^p with (4.20) determines the backstress and thus the kinematic hardening. As described in detail in Ch. 3 the finite strain post-yield behavior of thermoplastic polymers suggests to model the resistance against the orientation and stretch of molecular chains by the statistical network models of entropic rubber elasticity as originally proposed by Haward and Thackray (1968). Phenomena such as a pronounced Bauschinger effect or the recovery of a plastically deformed specimen to its original shape when heated up above the glass transition temperature gives rise to the assumption of aligned polymer chains in principal deformation directions in an entangled chain network.

6 The dependence of s_0 on μ would lead to a complicated relation between the temperature-dependent Young's modulus, respectively, the shear modulus and the inelastic shear strain rate, especially when the material enters the rubbery state. Thus, the athermal shear strength is assumed to be constant for all temperatures.

In this work the affine stretch[7], non-Gaussian statistical mechanics eight-chain network model by Arruda and Boyce (1993) is used since it yields good approximations of the strain hardening in uniaxial as well as plane strain tension states (Wu and van der Giessen, 1993). The idea behind this model is to consider eight chains of equal length in a unit cube along the diagonal directions representing an initially isotropic molecular network (Fig. 4.4). To compute the configurational entropy of the Langevin statistics and the resulting strain energy proportional to the change of entropy (Kuhn and Grün, 1942) it is assumed that these eight chains stretch affinely in the principal directions. The strain energy function $W^p = \rho_0 \Psi^p$ of the eight-chain model in terms of some mean chain stretch λ_C (defined later on) reads

$$W^p = C^R \lambda_L \left(\beta \lambda_C - \lambda_L \ln\left[\frac{\sinh \beta}{\beta} \right] \right) \tag{4.32}$$

with the rubber modulus C^R, the inverse of the Langevin function $\beta = \mathcal{L}^{-1}[\lambda_{CL}]$ and $\lambda_{CL} = \dfrac{\lambda_C}{\lambda_L}$ the ratio of the mean chain stretch and the limit stretch λ_L of a chain. The inverse Langevin function provides a highly non-linear force response of a chain at stretching. In Fig. 4.3 the inverse Langevin function β is plotted against the mean chain stretch. For increasing chain stretches β increases rapidly and tends to infinity when λ_C approaches the limit stretch ($\lim\limits_{\lambda_{CL} \to 1} \beta = \infty$). This corresponds to the finite extensibility of a chain. Since the inverse Langevin function cannot be represented in a closed-form it is appropriate to use the well performing Padé approximation (Cohen, 1991)

$$\beta = \lambda_{CL} \frac{3 - \lambda_{CL}^2}{1 - \lambda_{CL}^2} \quad . \tag{4.33}$$

[7] Affine stretch implies the direct connection between applied macroscopic deformation and the resulting molecular alignment in direction of the deformation.

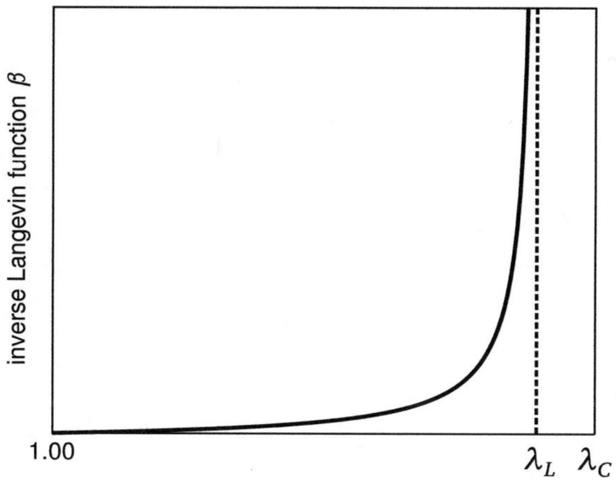

Figure 4.3.: Inverse Langevin function against mean chain stretch

In the eight-chain model, the mean chain stretch is defined by the ratio of the lengths of the end-to-end vectors $R = \frac{\sqrt{3}}{2}a$ of an undeformed chain and $r = \frac{a}{2}(\lambda_1^{p2} + \lambda_2^{p2} + \lambda_3^{p2})^{\frac{1}{2}}$ of a deformed chain in the unit cell of Fig. 4.4 of undeformed size a and deformed size $\lambda_i^p a$, respectively. Thus, the mean chain stretch is given by

$$\lambda_c = \frac{r}{R} = \frac{1}{\sqrt{3}}(\lambda_1^{p2} + \lambda_2^{p2} + \lambda_3^{p2})^{\frac{1}{2}} = \left(\frac{1}{3}\mathrm{tr}[\hat{\boldsymbol{b}}^p]\right)^{\frac{1}{2}} \quad . \tag{4.34}$$

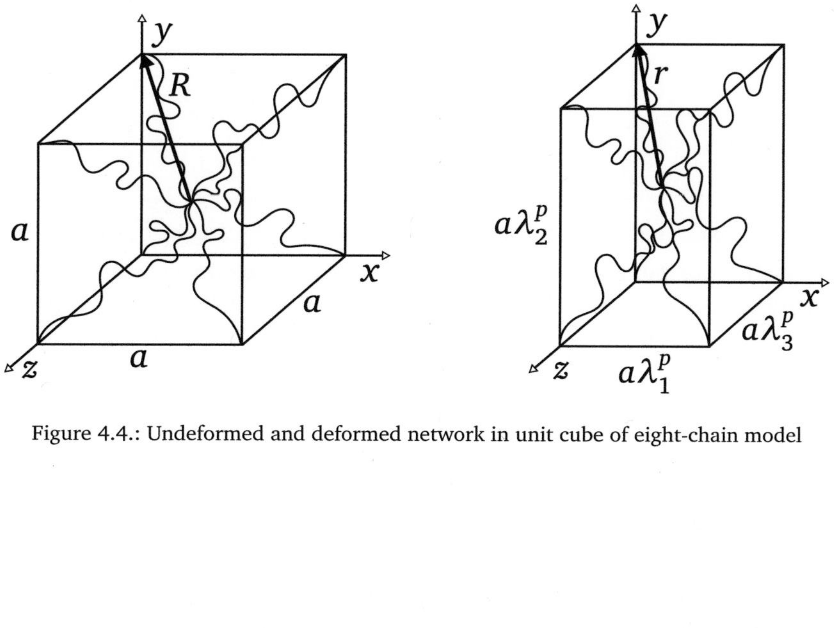

Figure 4.4.: Undeformed and deformed network in unit cube of eight-chain model

The limit stretch λ_L is derived from the freely jointed chain (FJC) model by defining a meaningful measure for the length of a chain (Ward and Sweeney, 2005). In modeling the molecular network of glassy polymers a "chain" is defined as the strand between two entanglements. It consists of an average number N of monomers, defined by rigid links of length l, so that the end-to-end vector of the chain is $\boldsymbol{R}_{FJC} = \sum_{i=1}^{N} \boldsymbol{l}_i$ where $\boldsymbol{l}_i$ is the vector of each rigid link (Fig. 4.5).

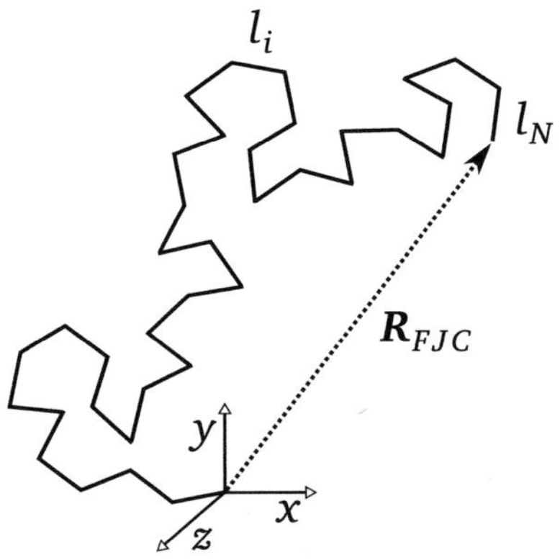

Figure 4.5.: Freely jointed chain

Since $\langle \boldsymbol{R}_{FJC} \rangle = \sum_{i=1}^{N} \langle \boldsymbol{l}_i \rangle = \boldsymbol{0}$ holds for the mean end-to-end vector due to free fluctuation it is not suited as a measure for the mean chain length[8]. The mean squared end-to-end length $\langle R_{FJC}^2 \rangle = \langle \boldsymbol{R}_{FJC} \cdot \boldsymbol{R}_{FJC} \rangle$ is used instead which yields

$$\langle R_{FJC}{}^2 \rangle = \sum_{i=1}^{N} \sum_{j=1}^{N} \langle \boldsymbol{l}_i \cdot \boldsymbol{l}_j \rangle = \sum_{i=1}^{N} \langle \boldsymbol{l}_i \cdot \boldsymbol{l}_i \rangle + 2 \sum_{i=1}^{N} \sum_{j>i}^{N} \langle \boldsymbol{l}_i \cdot \boldsymbol{l}_j \rangle = \sum_{i=1}^{N} l^2 = N l^2 \quad . \tag{4.35}$$

In the above equation, the relations $\langle \boldsymbol{l}_i \cdot \boldsymbol{l}_i \rangle = l^2$ and $\langle \boldsymbol{l}_i \cdot \boldsymbol{l}_j \rangle = l^2 \langle \cos \alpha_{ij} \rangle = 0$ are used. The latter one holds because for any possible angle in the range of $-\pi \leq \alpha_{ij} \leq \pi$ with equal probability we have $\langle \cos \alpha_{ij} \rangle = 0$. Hence, the mean end-to-end length of a freely jointed chain results in

$$R_{FJC} = \sqrt{\langle R_{FJC}^2 \rangle} = \sqrt{N} l \quad . \tag{4.36}$$

8 $\langle ... \rangle$ denotes the expectation value.

The length of a completely extended chain (strand between two entanglements) is $r_{FJC} = Nl$. The ratio of the lengths of the fully stretched and unstretched chain hence provides the limit stretch of a chain segment

$$\lambda_L = \frac{r_{FJC}}{R_{FJC}} = \sqrt{N} \quad . \tag{4.37}$$

In the theory of rubber elasticity, the modulus in (4.32) is determined by $C^R = nkT$ where n is the number of chains (i.e. strands between entanglements) per volume, k the Boltzmann constant and T the absolute temperature. Since the thus predicted increasing stiffness with increasing temperature is in contrast to experimental findings for the hardening behavior of amorphous thermoplastic polymers, C^R is taken here to be a temperature-independent material parameter. With the above definitions at hand, the deviatoric backstress tensor for the eight-chain model reads

$$\hat{\boldsymbol{\tau}}^b = \frac{C^R \beta}{3\lambda_{CL}} \hat{\boldsymbol{b}}^{p'} \quad . \tag{4.38}$$

in terms of the deviatoric part of the inelastic left Cauchy-Green tensor. Due to the behavior of the inverse Langevin function the backstress increases dramatically if λ_C approaches λ_L. Otherwise, for a mean chain stretch much smaller than the limit stretch the eight-chain model approximates a neo-Hookean material model.

Parameter fit

The material parameters are fitted to the experimental results of Ch. 3 for PMMA by using the optimization tool LS-OPT (LSTC, 2013a). For this purpose, the curve mapping procedure (LSTC, 2013b) is used which is basically an enhanced version of the commonly used mean squared error formulation. It leads to a much more stable computation of the error between the target (experimental) and the simulated curve especially when the target curve consists of steep, hysteretic or redundant parts (LSTC, 2013b). The main idea of this procedure is to map the points of the computed curve onto the target curve and then compute the resulting area between the two curves. This area serves as the mismatch, respectively, the error of the fitting which is to be minimized during the iterative optimization process. To get a much more precise error computation and thus a better fit, the mapping is done over a number of segmented parts of the two curves. For a much more detailed description it is referred to the manual of LS-OPT (LSTC, 2013b). The material parameter values resulting from the identification process are listed in Tab. 4.1. The Young's modulus as well as the strain hardening parameters of the eight-chain model

C^R and N are identified at room temperature (RT). The corresponding true stress-log. strain curves are presented in Fig. 4.6 in comparison with the experimental data. The model performs very well below the glass transition temperature; the onset of yielding and the hardening behavior are in good agreement with the experiments. However, for temperatures of 80 °C and above the yield stress and the hardening behavior are highly overestimated by the model.

Table 4.1.: Material parameter values for PMMA in the basic model for PMMA

intermolecular resistance		molecular network	
E [MPa]	2500	$N(T=293.15\text{ K})=N_{RT}$ [−]	2
ν [−]	0.4	$C^R(T=293.15\text{ K})=C^R_{RT}$ [MPa]	5
$\dot{\gamma}^p_0$ [s^{-1}]	$1.707\cdot 10^{25}$		
A [MPa K^{-1}]	131		
s_0 [MPa]	180		
α [−]	0.2		

4.2. Model extension beyond the glass transition temperature

In order to enhance the basic model to be able to capture the strong temperature dependency of amorphous polymers the inter- and intramolecular resistances against deformation are modified in the following. In case of the intermolecular resistance, the model is modified in terms of Young's modulus and the yield strength as discussed below, while two different extensions of the intramolecular resistance are separately analyzed in sections Sec. 4.2.1 and Sec. 4.2.2.

Firstly, Young's modulus is taken to be temperature-dependent[9] according to the experimental results in Fig. 3.24a. In the glassy state the modulus decreases linearly until the glass transition region begins. The onset of this region is indicated by $T_r = \mathrm{T_g} - \Delta\mathrm{T_g}/2$

[9] Here, Young's modulus is temperature-dependent only. Poisson's ratio is taken to be constant for simplicity. In contrast to that, Dupaix and Boyce (2007) modeled Poisson's ratio to be temperature-dependent; the polymer is compressible ($\nu = 0.33$) in the glassy state below the glass transition temperature $\mathrm{T_g}$ and incompressible ($\nu = 0.5$) in the rubbery state above $\mathrm{T_g}$.

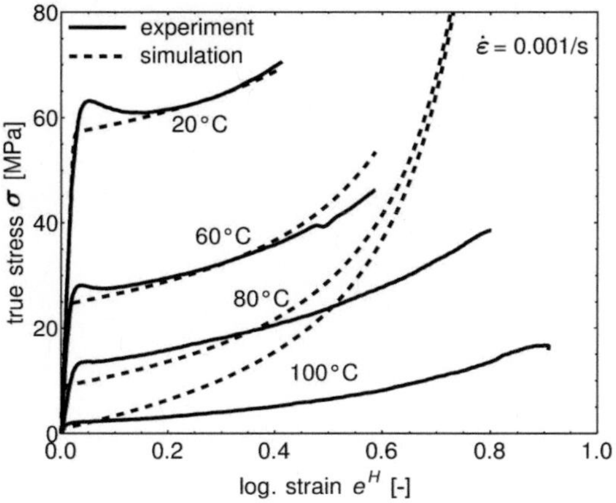

(a)

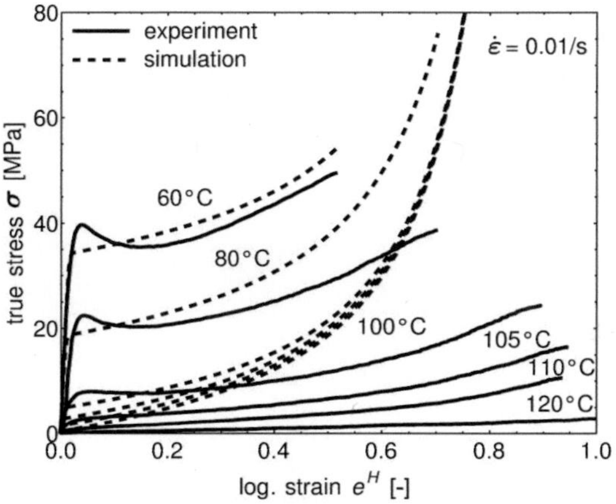

(b)

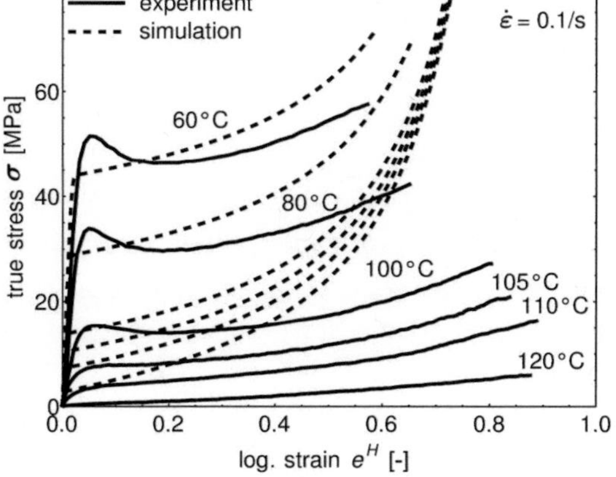

(c)

Figure 4.6.: Comparison of experimental data for PMMA and model response where ΔT_g gives the size of the glass transition region in which the drop of the stiffness takes place. This drop is determined by a hyperbolic tangent function (cf. Fig. 4.7, Dupaix and Boyce (2007)) at temperatures above T_r. Accordingly, Young's modulus is given by

$$E(T) = \begin{cases} E_{g1} - E_{g2} T & , \quad \text{if } T < T_r \\ \frac{1}{2}(E_{gr} + E_{r1}) - \frac{1}{2}(E_{gr} - E_{r1}) \tanh\left[\frac{5}{\Delta T_g}(T - T_g)\right] - E_{r2}(T - T_g) \ , & \text{if } T \geq T_r \end{cases} \tag{4.39}$$

with the material parameters E_{g1}, E_{g2}, E_{r1}, E_{r2} and $E_{gr} = E_{g1} - E_{g2} T_r$. E_{r2} represents the remaining stiffness per temperature after the drop which further decreases linearly for increasing temperatures[10]. This temperature-dependent Young's modulus is incorporated into the model through the Lamé parameters appearing in the strain energy function (4.23).

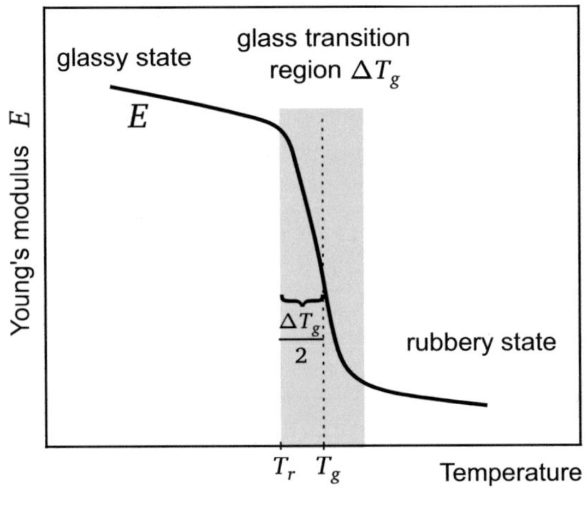

Figure 4.7.: Modelling of temperature-dependent Young's modulus

Secondly, the resistance against the onset of yield, namely the inelastic shear strain rate is modified as well. Although the temperature dependency is already included by the Arrhenius-type equation (4.29) this is not sufficient to capture the variation of the yield strength at temperatures around T_g (Fig. 4.6). For this reason (4.31) is modified to

$$\dot{\gamma}^p = \dot{\gamma}_0^p \left(\exp\left[\tilde{A}(T) \left(\|\boldsymbol{\Sigma}^{*\prime}\| - s\right)\right] - \exp\left[-\tilde{A}(T)\, s\right] \right) \quad . \tag{4.40}$$

[10] The constraint of $E(T) > 0$ must be enforced by the material parameters and the regarded temperature range, respectively.

The temperature-dependent function $\tilde{A}(T)$ is introduced in a phenomenological manner and reads as

$$\tilde{A}(T) = A_0 \exp(-A_T T) \tag{4.41}$$

with the material parameters A_0 and A_T. This additional function may be interpreted to capture a second relaxation process (α relaxation at T_g) to account for the transition to the rubbery state of the material without introducing explicit constitutive equations for this additional relaxation. With more complex functions for the inelastic shear rate this was also done by Dooling et al. (2002b), Richeton et al. (2007b), Srivastava et al. (2010), for example.

The above introduced material parameters for the intermolecular resistance are determined from a parameter fit (see Sec. 4.1.2) in the region of small strains ($e^H \leq 0.05$). These parameter values are given in Tab. 4.2 and are used in the following.

Table 4.2.: Material parameter values of intermolecular resistance modification for PMMA

Young's modulus		shear yielding	
T_g [K]	378.15	$\dot{\gamma}_0^p[\mathrm{s}^{-1}]$	$1 \cdot 10^{26}$
ΔT_g [K]	15	s_0 [MPa]	180
E_{g1} [MPa]	9096	A_0 [MPa^{-1}]	1.213
E_{g2} [MPa $^{-1}$]	22.5	A_T [K^{-1}]	0.00323
E_{r1} [MPa]	100	α [−]	0.2
E_{r2} [MPa K^{-1}]	4.15		

To incorporate a temperature dependence of the molecular network at large strains, two different molecular network models are presented in the following two sections.

4.2.1. Entanglement dissociation model of Raha and Bowden

In this section the resistance of the molecular network against deformation, hence the eight-chain model for the backstress, is taken to be temperature-dependent by incorporating the "entanglement dissociation model" of Raha and Bowden (1972). In their work they studied the evolution of birefringence in PMMA samples under inelastic deformation. Based on the results they proposed a model for a temperature-dependent variation of the chain or entanglement density, respectively, where a chain is defined as a segment between two entanglements. The idea of their model is that the physically entangled molecular network evolves with increasing temperature such that secondary valence bonds (weak entanglements) dissociate with increasing temperature. The proposed relation for the decreasing chain density with increasing temperature was slightly modified by Arruda et al. (1995) and reads

$$n(T) = B - D \exp\left[-\frac{E_a}{R\,T}\right] \tag{4.42}$$

where B describes the strong, temperature-independent network consisting of permanent entanglements. The second term in (4.42) represents with D the weak, temperature-dependent portion of the network which dissociates with increasing temperature. R is the universal gas constant[11] and E_a is the thermal dissociation energy.

Arruda et al. (1995) incorporated the relation for the chain density (4.42) into the eight-chain model to capture the temperature-dependent strain hardening.With a decreasing number of entanglements in the network, the number of rigid links N between entanglements must increase due to the conservation of mass. Thus, the relation

$$N(T)\; n(T) = \text{const.} \tag{4.43}$$

must hold. The consequence of (4.43) is a temperature dependence of the number of rigid links $N(T)$ and according to that the relation $N(T)n(T) = N_{RT}n_{RT}$ must hold where the index RT indicates the "room temperature". Hence, the temperature-dependent average number of rigid links between entanglements is given by

$$N(T) = N_{RT}\frac{n_{RT}}{n(T)} \quad . \tag{4.44}$$

Thus, the limit stretch $\lambda_L(T) = \sqrt{N(T)}$ of a chain between entanglements increases with increasing temperature. This temperature-dependent limit stretch is directly incorporated in the eight-chain model (4.38) so that the strain hardening decreases with increasing temperature. By incorporating the temperature-dependent chain density $n(T)$ as defined in (4.42) with fitted values for the material parameters B, D and E_a at elevated temperatures into the rubbery modulus $C^R = n(T)kT$, it yields $C^R_{RT} \neq n_{RT}kT_{RT}$, where C^R_{RT} is gained from the fitting procedure of Sec. 4.1.2. Therefore, in this work a scaled rubbery modulus C^{R*} is introduced which is gained as

$$C^{R*}(T) = C^R(T)\frac{C^R_{RT}}{n_{RT}kT_{RT}} = n(T)kT\frac{C^R_{RT}}{n_{RT}kT_{RT}} = C^R_{RT}\frac{n(T)T}{n_{RT}\,T_{RT}} \tag{4.45}$$

so that $C^{R*}(T = RT) = C^R_{RT}$ holds.

[11] The value of the universal gas constant is $R = 0.001695$ kcal/mol K

Parameter fit

A material parameter fit (see Sec. 4.1.2) for PMMA yields the values given in Tab. 4.3. The uniaxial tensile true stress-log. strain response of the model compared to the experimental data is shown in Fig. 4.8. A very good agreement with the experiments is achieved for all temperatures and strain rates. However, since the model bases on the thermoelastic relation (4.42) it does not capture the irreversibility of entanglement dissociation.

Table 4.3.: Material parameter values for PMMA in entanglement dissociation model

entanglement dissociation model	
B [mm^{-3}]	$2.9547 \cdot 10^{18}$
D [mm^{-3}]	$1.20087 \cdot 10^{22}$
E_a [kcal mol^{-1}]	6.543

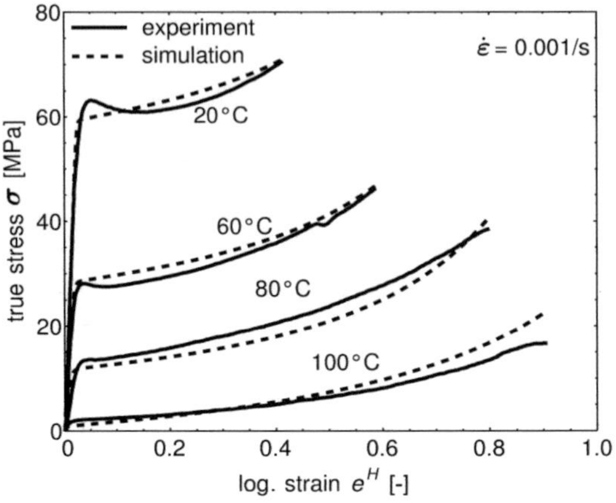

(a)

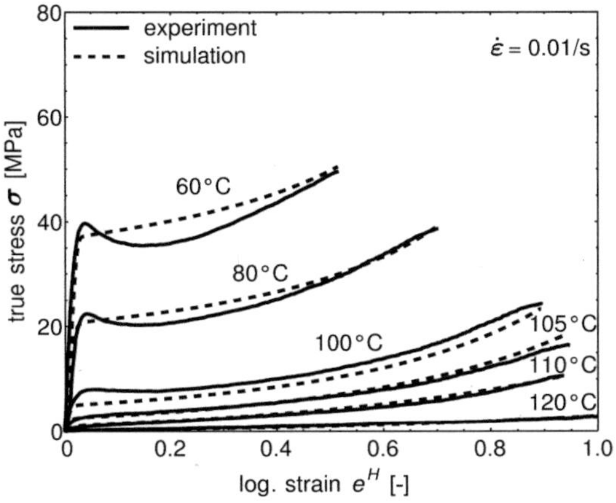

(b)

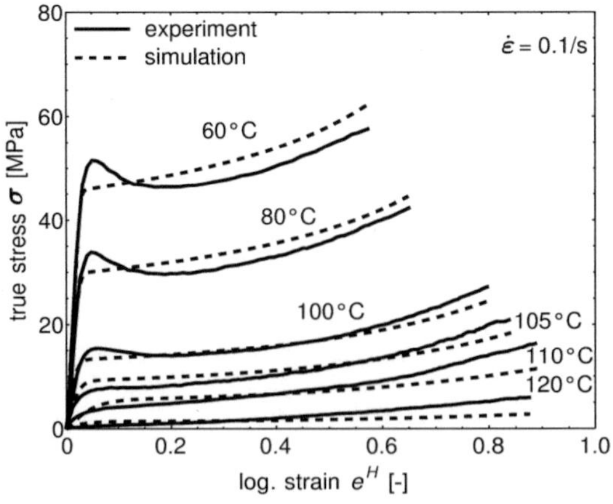

(c)

Figure 4.8.: Comparison of experimental data for PMMA and response of entanglement dissociation model

4.2.2. Molecular relaxation by reptation

In this section the basic model of Sec. 4.1.2 is enhanced to overcome the drawbacks of the model discussed in the previous section. For this purpose, the part of the model representing the molecular network is extended to be temperature- and strain rate-dependent. Again, a rheological model serves as an illustration of the constitutive model (Fig. 4.9). In contrast to Fig. 4.1 an additional dashpot is incorporated into the molecular network part (N) in order to represent the temperature- and strain rate-dependent relaxation mechanism.

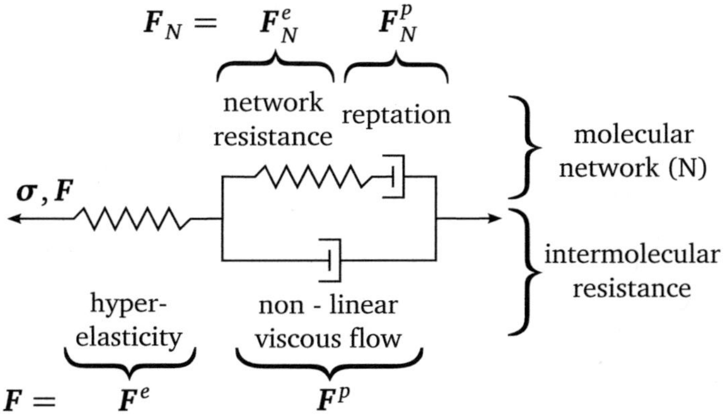

Figure 4.9.: Rheological model including molecular relaxation by reptation

The mechanism of polymer chain relaxation due to reptation is responsible for the time and temperature dependence of hardening. Reptation denotes the sliding of chains in an entangled network under load at elevated temperatures through a tube-like path which is formed by entanglements of surrounding chains (de Gennes (1971), de Gennes (1979), Doi and Edwards (1978)). This mechanism is the background of previous works modeling the strain rate-dependent behavior of elastomers (Bergström and Boyce, 1998) or the temperature and strain rate dependence of the strain hardening at elevated temperatures of thermoplastic polymers, e.g. Dooling et al. (2002b), Dupaix and Boyce (2007), Boyce et al. (2000).

Kinematics

Based on the kinematics of the basic model (Sec. 4.1.1), the kinematics of the present model is extended to account for the additional relaxation mechanism. Therefore, the formerly inelastic part of the deformation gradient $\boldsymbol{F}^p$ is split in two parts representing the network deformation and relaxation as

$$\boldsymbol{F}^p = \boldsymbol{F}_N = \boldsymbol{F}^e_N \boldsymbol{F}^p_N \tag{4.46}$$

with the elastic network part $\boldsymbol{F}^e_N$ which describes the alignment of the molecular chains. The inelastic network part $\boldsymbol{F}^p_N$ represents the molecular reptation[12]. This is equivalent to introducing a second intermediate configuration $\tilde{\mathcal{B}}$. To clarify this concept of two virtual intermediate configurations the situation is sketched in Fig. 4.10.

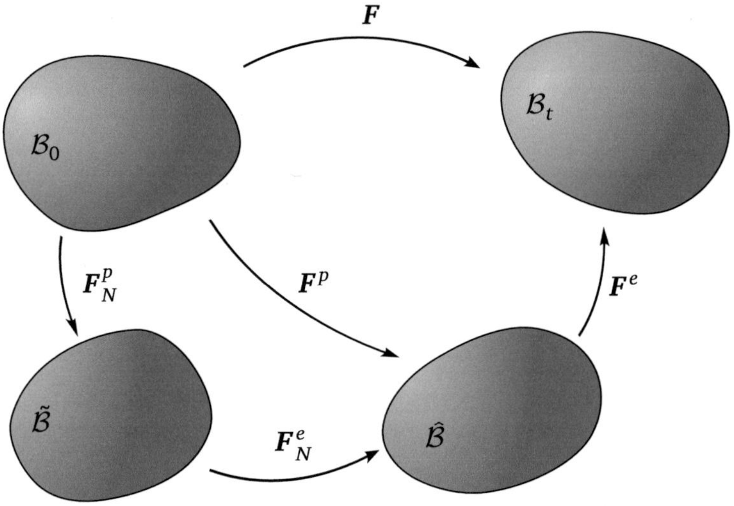

Figure 4.10.: Additional split of the inelastic deformation gradient resulting in a second intermediate configuration $\tilde{\mathcal{B}}$

With these definitions at hand, the spatial velocity gradient may be written as

$$\begin{aligned} \boldsymbol{l} &= \dot{\boldsymbol{F}}\boldsymbol{F}^{-1} = \dot{\boldsymbol{F}}^e\boldsymbol{F}^{e-1} + \boldsymbol{F}^e\dot{\boldsymbol{F}}^e_N\boldsymbol{F}^{e-1}_N\boldsymbol{F}^{e-1} + \boldsymbol{F}^e\boldsymbol{F}^e_N\dot{\boldsymbol{F}}^p_N\boldsymbol{F}^{p-1}_N\boldsymbol{F}^{e-1}_N\boldsymbol{F}^{e-1} \\ &= \boldsymbol{l}^e + \dot{\boldsymbol{F}}^e\hat{\boldsymbol{l}}^e_N\boldsymbol{F}^{e-1} + \boldsymbol{F}^e\boldsymbol{F}^e_N\tilde{\boldsymbol{l}}^p_N\boldsymbol{F}^{e-1}_N\boldsymbol{F}^{e-1} & (4.47) \\ &= \boldsymbol{l}^e + \boldsymbol{l}^e_N + \boldsymbol{F}^e\hat{\boldsymbol{l}}^p_N\boldsymbol{F}^{e-1} = \boldsymbol{l}^e + \boldsymbol{l}^e_N + \boldsymbol{l}^p_N = \boldsymbol{l}^e + \boldsymbol{l}^p & (4.48) \end{aligned}$$

where $\tilde{\boldsymbol{l}}^p_N$ is the inelastic network velocity gradient in the intermediate configuration $\tilde{\mathcal{B}}$ and $\hat{\boldsymbol{l}}^e_N$ denotes the elastic network velocity gradient in the intermediate configuration $\hat{\mathcal{B}}$. Utilizing the same arguments as in Sec. 4.1.1 the inelastic spin tensors $\hat{\boldsymbol{w}}^p$ and $\tilde{\boldsymbol{w}}^p_N$ are

[12] Quantities describing the molecular network are indicated by the subscript N.

set to zero while the inelastic rate of deformation tensors $\hat{\boldsymbol{d}}^p$ and $\tilde{\boldsymbol{d}}^p_N$ are constitutively prescribed. This will be done in the next section by introducing additional constitutive equations for the relaxation of molecular chains.

Constitutive equations

As in Sec. 4.1.2, a structure for the constitutive equations in line with the second law of thermodynamics is given first. The free energy density is additively split into an elastic intermolecular resistance part Ψ^e and a network resistance part Ψ^e_N. The stored elastic energy than depends on the elastic right Cauchy-Green tensor $\hat{\boldsymbol{C}}^e = \boldsymbol{F}^{eT}\boldsymbol{F}^e$, the left Cauchy-Green tensor $\hat{\boldsymbol{b}}^e_N = \boldsymbol{F}^e_N\boldsymbol{F}^{eT}_N$ in the intermediate configuration $\hat{\mathcal{B}}$ related to the elastic network deformation and the temperature T. Hence, the free energy density reads

$$\Psi = \Psi^\dagger(\hat{\boldsymbol{C}}^e, \hat{\boldsymbol{b}}^e_N, T) = \Psi^e(\hat{\boldsymbol{C}}^e, T) + \Psi^e_N(\hat{\boldsymbol{b}}^e_N, T) \tag{4.49}$$

and its time derivative is given by

$$\dot{\Psi} = \frac{\partial \Psi^e}{\partial \hat{\boldsymbol{C}}^e} : \dot{\hat{\boldsymbol{C}}}^e + \frac{\partial \Psi^e}{\partial T}\dot{T} + \frac{\partial \Psi^e_N}{\partial \hat{\boldsymbol{b}}^e_N} : \dot{\hat{\boldsymbol{b}}}^e_N + \frac{\partial \Psi^e_N}{\partial T}\dot{T} \quad . \tag{4.50}$$

Applying kinematic relations as in Sec. 4.1.2 and using additionally $\dot{\hat{\boldsymbol{b}}}^e_N = \hat{\boldsymbol{l}}^e_N\hat{\boldsymbol{b}}^e_N + \hat{\boldsymbol{b}}^e_N\hat{\boldsymbol{l}}^{eT}_N$, $\hat{\boldsymbol{d}}^p = \hat{\boldsymbol{d}}^e_N + \hat{\boldsymbol{d}}^p_N$ with $\hat{\boldsymbol{d}}^p_N = sym[\boldsymbol{F}^e_N\tilde{\boldsymbol{l}}^p_N\boldsymbol{F}^{e-1}_N]$ and inserting these into (2.93) one gets

$$\begin{aligned}(\boldsymbol{\tau} - 2\rho_0\boldsymbol{F}^e\frac{\partial \Psi^e}{\partial \hat{\boldsymbol{C}}^e}\boldsymbol{F}^{eT}) : \boldsymbol{d} + (2\rho_0\boldsymbol{F}^{eT}\boldsymbol{F}^e\frac{\partial \Psi^e}{\partial \hat{\boldsymbol{C}}^e}\boldsymbol{F}^{eT}\boldsymbol{F}^{e-T} - 2\rho_0\hat{\boldsymbol{b}}^e_N\frac{\partial \Psi^e_N}{\partial \hat{\boldsymbol{b}}^e_N}) : \hat{\boldsymbol{d}}^p \\ +2\rho_0\boldsymbol{F}^{eT}_N\hat{\boldsymbol{b}}^e_N\frac{\partial \Psi^e_N}{\partial \hat{\boldsymbol{b}}^e_N}\boldsymbol{F}^{e-T}_N : \tilde{\boldsymbol{d}}^p_N - (\frac{\partial \Psi^e}{\partial T} + \frac{\partial \Psi^e_N}{\partial T} + s)\rho_0\dot{T} \geq 0 \quad .\end{aligned} \tag{4.51}$$

Again, this yields the potential relation for the entropy $s = -(\frac{\partial \Psi^e}{\partial T} + \frac{\partial \Psi^e_N}{\partial T})$ and the Kirchhoff stress tensor $\boldsymbol{\tau}$ as defined before in (4.18). The driving stress tensor $\boldsymbol{\Sigma}^*$ defined in (4.21) is given through the Mandel stress tensor $\boldsymbol{\Sigma}$ (4.19) and the backstress tensor in the intermediate configuration $\hat{\mathcal{B}}$

$$\hat{\boldsymbol{\tau}}^b = 2\rho_0\hat{\boldsymbol{b}}^e_N\frac{\partial \Psi^e_N}{\partial \hat{\boldsymbol{b}}^e_N} \quad . \tag{4.52}$$

The backstress tensor $\tilde{\boldsymbol{\tau}}^b = \boldsymbol{F}^{eT}_N\hat{\boldsymbol{\tau}}^b\boldsymbol{F}^{e-T}_N$ with regards to the intermediate configuration $\tilde{\mathcal{B}}$ is obtained by a pull-back operation and is work conjugated to the inelastic network rate of deformation tensor $\tilde{\boldsymbol{d}}_N^p$. Hence, the remaining inequality (4.51) and internal dissipation reads

$$\mathcal{D}_{int} = \boldsymbol{\Sigma}^* : \hat{\boldsymbol{d}}^p + \tilde{\boldsymbol{\tau}}^b : \tilde{\boldsymbol{d}}_N^p \geq 0 \quad . \tag{4.53}$$

The inelastic rate of deformation tensor $\hat{\boldsymbol{d}}^p$ is defined through the flow rule (4.27) with the inelastic shear strain rate $\dot{\gamma}^p$ defined in (4.40). As shown before, this fulfills the restriction of positive internal dissipation, cf. (4.28). So, the second term in (4.53) must be equal or greater than zero as well. For this reason, an evolution equation similar to the flow rule (4.27) is chosen for the inelastic network rate of deformation

$$\tilde{\boldsymbol{d}}_N^p = \dot{\gamma}_N^p \frac{\tilde{\boldsymbol{\tau}}^{b\prime}}{\|\tilde{\boldsymbol{\tau}}^b\|} \tag{4.54}$$

which depends on the rate of molecular relaxation $\dot{\gamma}_N^p$ and the deviatoric backstress $\tilde{\boldsymbol{\tau}}^b$. A result of inserting this equation in the internal dissipation (4.53) is that the requirement $\dot{\gamma}_N^p \geq 0$ must hold. Same strain energy functions $W^e = \rho_0 \Psi^e$ (see (4.23)) and $W_N^e = \rho_0 \Psi_N^e$ (see (4.32)) are applied as in Sec. 4.1.2 so that the Kirchhoff stress is determined by the hyperelastic neo-Hooke model and the deviatoric backstress by the eight-chain model which depends on the deviatoric elastic network left Cauchy-Green tensor

$$\hat{\boldsymbol{\tau}}^b = \frac{C^R \beta}{3\lambda_{CL}} \hat{\boldsymbol{b}}_N^{e\prime} \tag{4.55}$$

with the ratio $\lambda_{CL} = \dfrac{\lambda_C}{\lambda_L}$ of the the mean chain stretch

$$\lambda_C = \left(\frac{1}{3}\mathrm{tr}[\hat{\boldsymbol{b}}_N^e]\right)^{\frac{1}{2}} \tag{4.56}$$

and the limit stretch λ_L.

The only remaining undetermined quantity is the rate of relaxation $\dot{\gamma}_N^p$ for which a phenomenological relation is suggested. It is therefore obvious to choose a function which depends on the temperature. It is also necessary to consider the network stress (backstress) to include the rate dependency and furthermore the network deformation state must be taken into account to control the cessation of the relaxation process at large strains. In total, the rate of relaxation is taken to consist of three functions

$$\dot{\gamma}_N^p = f_T(T) f_{\tilde{\tau}^b}(\tilde{\boldsymbol{\tau}}^b) f_{\lambda_N}(\lambda_N) \tag{4.57}$$

where $\lambda_N = \left(\frac{1}{3}\text{tr}[\hat{\boldsymbol{b}}^p]\right)^{\frac{1}{2}}$ is the total network stretch. To capture the temperature dependency of the relaxation an Arrhenius type exponential relation is chosen as proposed by Bergström and Boyce (1998)

$$f_T(T) = \dot{\gamma}^p_{NO} \exp\left[-\frac{A_{NT}}{T - 273.15}\right] \tag{4.58}$$

with two material parameters $\dot{\gamma}^p_{NO}$ and A_{NT}. The rate dependency is included via the power-law

$$f_{\tilde{\tau}^b}(\tilde{\tau}^b) = \left(\frac{\|\tilde{\tau}^b\|}{\tilde{\tau}^b_0}\right)^{\kappa} \tag{4.59}$$

with the material parameters κ and $\tilde{\tau}^b_0$. The cessation of the relaxation process is described by the function

$$f_{\lambda_N}(\lambda_N) = \begin{cases} \left(\dfrac{\lambda_N - \lambda^L_N}{1 - \lambda^L_N}\right)^{\xi} & , \quad \text{if } \lambda_N \leq \lambda^L_N \\ 0 & , \quad \text{if } \lambda_N > \lambda^L_N \end{cases} \tag{4.60}$$

which is similar to those proposed by Adams et al. (2000) or Dupaix and Boyce (2007). The parameter λ^L_N represents the relaxation limit stretch of the network at which the relaxation is completely finished. This type of function requires the usage of the total network stretch λ_N as the controlling measure for the cessation of relaxation. If the chain stretch is chosen instead only, the stress relaxes to zero at high temperatures because no chain stretch is present at the beginning and could not be developed at further deformation due to the continuing relaxation. While the dependency of (4.60) gives good results at very high temperatures, the hardening at lower temperatures is overestimated. For this reason, the relaxation limit stretch λ^L_N is taken to be temperature-dependent as well with the exponential function

$$\lambda^L_N(T) = 1 + \lambda^L_{NO} \exp[\lambda^L_{NT}\ T] \tag{4.61}$$

where λ^L_{NO} and λ^L_{NT} are material parameters. These parameters are chosen in a manner so that for the function $\lambda^L_N(T)$ the relation

$$\lambda^L_N(T) = \begin{cases} \approx 1 & \text{if } T < \text{T}_\text{g} \\ > 1 & \text{if } T \geq \text{T}_\text{g} \end{cases} \tag{4.62}$$

holds and the influence below T_g vanishes.

All together, the rate of relaxation is given by

$$\dot{\gamma}_N^p = \begin{cases} \dot{\gamma}_{N0}^p \exp\left[-\frac{A_{NT}}{T}\right]\left(\frac{\lambda_N - \lambda_N^L}{1-\lambda_N^L}\right)^{\xi}\left(\frac{\|\tilde{\tau}^b\|}{\tilde{\tau}_0^b}\right)^{\kappa} & , \text{ if } \lambda_N \leq \lambda_N^L \\ 0 & , \text{ if } \lambda_N > \lambda_N^L \end{cases} \tag{4.63}$$

and is characterized by seven material parameters. The form of this function ensures that $\dot{\gamma}_N^p \geq 0$ holds for thermodynamic consistency if $\dot{\gamma}_{N0}^p$ is positive.

Parameter fit

The parameter fitting is done analogously to Sec. 4.1.2. As before the parameters for the intermolecular resistance determined from the small strain regime are used (Tab. 4.2). The additional parameters for PMMA introduced here are listed in Tab. 4.4. Note, that no modifications in the eight-chain model are done and the parameters N and C^R are used as obtained at room temperature. Dupaix and Boyce (2007) in contrast, fitted N and C^R at temperatures close to T_g to capture the reduced hardening. A complete list of all used parameters is given in the App. A.1.

The response of the in the present section extended model is shown in terms of true stress-log. strain curves in Fig. 4.11 and is in good agreement with the experimental data. At moderate temperatures and higher strain rates the model somewhat overestimates the strain hardening of the experiment. This behavior can be explained by adiabatic heating occurring in tests at higher strain rates which is not considered in the model.

Table 4.4.: Material parameter values for PMMA in molecular relaxation model

molecular network reptation model	
$\dot{\gamma}_{N0}^p$ [s^{-1}]	$3.3 \cdot 10^{23}$
A_{NT} [K]	20541
κ [−]	6.756
λ_{N0}^L [−]	$3.9 \cdot 10^{-11}$
λ_{NT}^L [K^{-1}]	0.06306
ξ [−]	55
$\tilde{\tau}_0^b$ [MPa]	1

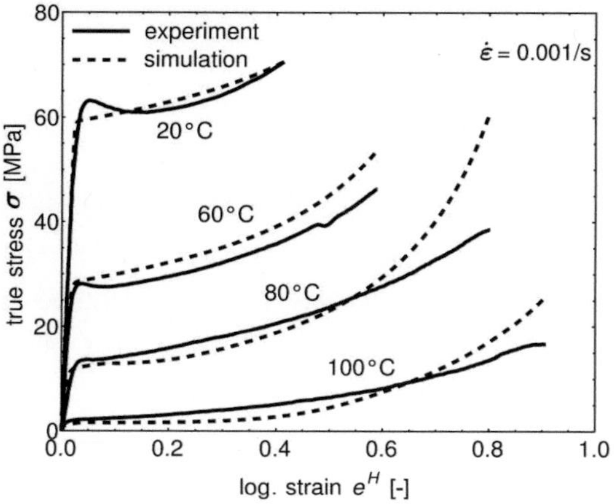

(a)

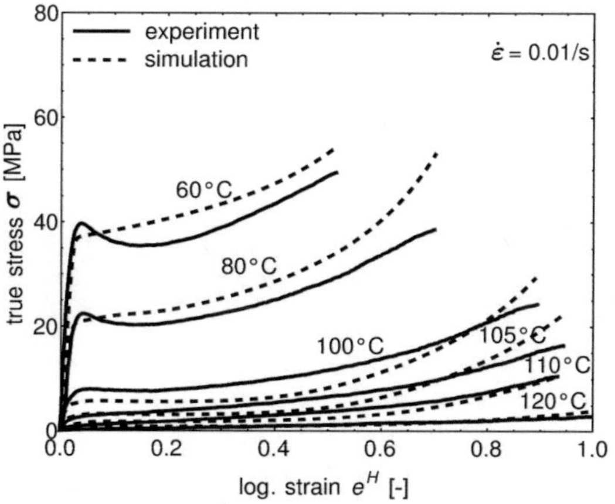

(b)

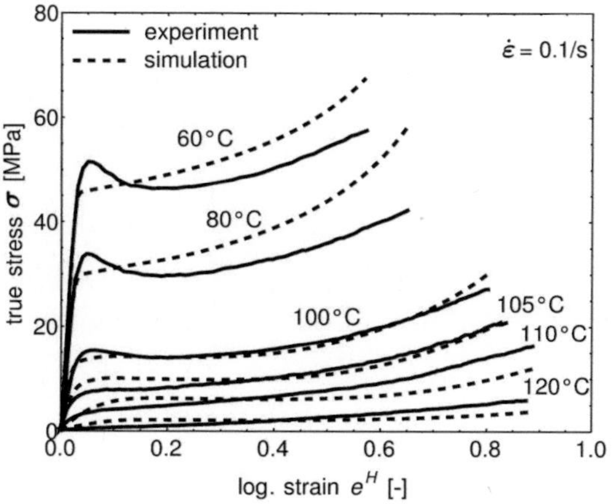

(c)

Figure 4.11.: Comparison of experimental data for PMMA and response of molecular relaxation model

4.2.3. Comparison of models for thermomechanical loading history

In order to analyze the response of the "basic" model (Sec. 4.1.2), the "entanglement dissociation" model (Sec. 4.2.1) and the "molecular relaxation by reptation" model (Sec. 4.2.2) under thermomechanical loading, the different material models are subjected to a fictitious deformation-temperature cycle. The applied thermomechanical loading history is similar to a forming process with an additional subsequent reheating. In the first step, the material is stretched uniformly up to $e_{11}^{H} = 0.7$ at a constant temperature of 110 °C during a time interval of 1000 s. In the second step, the deformation is held constant at the same constant temperature and time period. In the third step the temperature is decreased linearly to 20 °C over a time period of 2000 s at fixed deformation. In the fourth step a stress free state is attained; again at a constant temperature of 20 °C and a duration of 1000 s. In the fifth step the material is reheated to 110 °C in 1000 s and the temperature is held constant for 500 s. The deformation is still unconstrained in this step. The "process" steps of the thermomechanical loading are summarized in Tab. 4.5. The three model responses are shown in Fig. 4.12. For each case the Cauchy stress, backstress (both scaled by 3 s_0), log. strain and temperature (scaled by 110 °C) are plotted.

Table 4.5.: Steps of thermomechanical loading

#	process step	duration [s]	temperature [°C]
1.	uniform deformation at constant temperature	1000	110
2.	constrained deformation at constant temperature	1000	110
3.	constrained deformation at decreasing temperature	2000	110 − 20
4.	release to zero stress	1000	20
5.	reheating at unconstrained deformation	1500	20 − 110

In Fig. 4.12a, the response of the basic model is presented. Due to the missing temperature dependence of the strain hardening the stress response after the first step is far too high, cf. Fig. 4.6. In step 2 and 3, the stress remains constant since thermoelasticity is not considered in the model. In step 4 the material is released so that the Cauchy stress vanishes; the backstress is still present at the end of this step. At increasing temperature in the fifth step the relaxing backstress causes a re-deformation to zero strain (and zero backstress) which starts at approximately 50 °C.

The response of the "entanglement dissociation" model is shown in Fig. 4.12b. The reached Cauchy stress and backstress are much lower than in the basic model due to the temperature dependence of Young's modulus and the hardening behavior and due to the modified resistance against the onset of yield. In the second step no change of the load parameters occurred so that no stress change is present. In the third step the Cauchy stress as well as the backstress increase at constrained deformation. With decreasing temperature the Cauchy stress increases due to the increasing backstress and Young's modulus. The rise of the backstress is caused by the "stiffening spring" of the thermoelastic (enhanced) eight-chain model with the relation for the increasing chain density $n(T)$ (4.42) at decreasing temperature. In terms of the model, this is equivalent to formations of new entanglements. But in the constraint deformation state, the molecular network is still stretched and the formation of new entanglements in this configuration is not realistic. The model does not capture the irreversibility of the entanglement dissociation and thus predicts a far too high backstress stored in the material when the deformation is constrained and the temperature is decreased. For this reason the model is not suitable for a complex thermomechanical loading histories.

The response of the "molecular relaxation by reptation" model is shown in Fig. 4.12c. It is clearly observable that the Cauchy and the backstress in the first step are much lower than in the "entanglement dissociation" model due to the rate dependence of the hardening behavior. In the second step a small portion of the stresses relax already due to the molecular relaxation. The Cauchy stress increases in the third step due to increasing Young's modulus while the backstress remains constant which is true for the fourth step as well. This is a major difference to the entanglement dissociation model which predicts an unrealistic backstress raise in the third step due to the temperature-dependent eight-chain model discussed above. In the reheating (fifth) step the backstress relaxes and deforms the material but much less compared to the both previously discussed models. This is due to the much lower backstress stored in the material when stretching takes place at low rates and high temperatures.

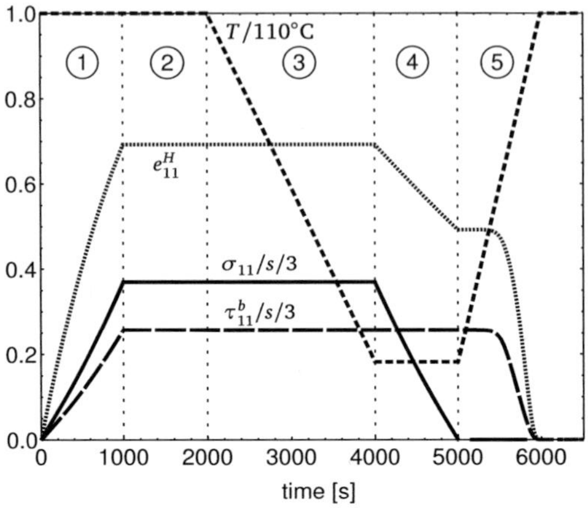

(a) basic model

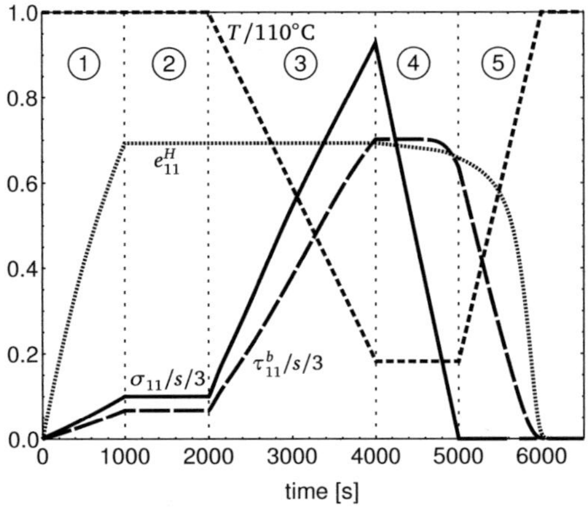

(b) entanglement dissociation model

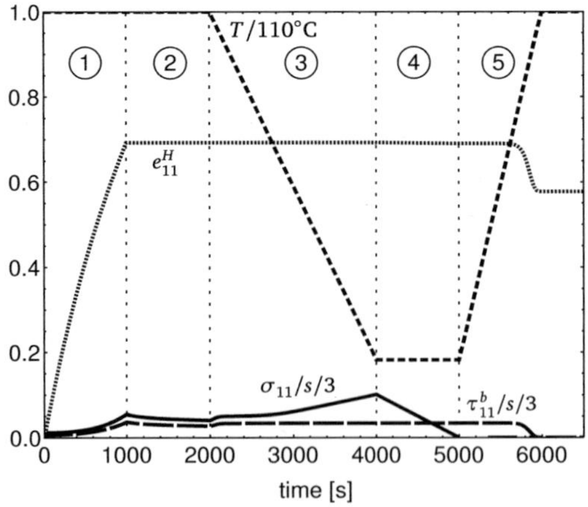

(c) molecular relaxation model

Figure 4.12.: Comparison of model responses to a thermomechanical loading history according Tab.4.5

This example shows that the "molecular relaxation by reptation" model is much more suitable for using in simulations of applications where such a thermomechanical load is involved, e.g. thermoforming. For this reason this model is used for thermoforming simulations of Ch. 5.

4.3. Model extension to account for initial orientation

Owing to preceding manufacturing processes such as injection molding or extrusion, thermoplastic polymer materials may exhibit a "frozen-in" pre-stretch one orientation of the molecular network (cf. Sec. 3.1.2). This pre-orientation may strongly affect the mechanical behavior in a subsequent deformation process, e.g. thermoforming, and - in order to avoid consideration of the entire thermomechanical history starting from a virgin material - needs to be accounted for in corresponding numerical simulations. The molecular pre-orientation is correlated to the measurable birefringence of the material (e.g. Kahar et al. (1978), De Focatiis and Buckley (2011)) and thus can be used as an available input quantity in a material model. For this purpose, the constitutive model is extended to account for an initial pre-stretch and orientation of the molecular network. In the model description the assumption is made that the pre-stretched network affects only the inelastic (hardening) mechanical behavior while the influence on the small strain elastic behavior can be neglected. Therefore the molecular network is modified through the backstress tensor introduced in (4.38) which originally depends on the network stretch with respect to an initial isotropic network. The incorporation of the initial material state through a pre-stretch tensor is discussed in the following section.

4.3.1. Incorporation of molecular pre-stretch tensor

To model a pre-stretched network Boyce et al. (1989a) introduced an initial network stretch tensor $\boldsymbol{V}_N^i$ which contains information about the pre-deformation history of the network. The eigenvalues $\lambda_{N\alpha}^i(\alpha = 1, 2, 3)$ of $\boldsymbol{V}_N^i$ represent the pre-stretch of the molecular network with the constraint $\det[\boldsymbol{V}_N^i] = 1$ according to the isochoric network deformation. Thus, the initial stretch as well as the following current deformation must be considered in the computation of the backstress. For this purpose a *network state deformation gradient* $\boldsymbol{F}_S^p$ is introduced

$$\boldsymbol{F}_S^p = \boldsymbol{F}^p \boldsymbol{V}_N^i \tag{4.64}$$

consisting of the current inelastic part of the deformation gradient and the initial network stretch tensor. For an initially isotropic molecular network, this tensor reduces to the inelastic deformation gradient only ($\lambda^i_{N1} = \lambda^i_{N2} = \lambda^i_{N3} = 1$). In Fig. 4.13 the different portions of the deformation with a pre-stretched network in the reference configuration $\mathcal{B}_0$ is schematically depicted. Starting from an initially isotropic ("virgin") material configuration, the network is stretched and oriented (e.g. due to a manufacturing process) as represented by $\boldsymbol{V}^i_N$. The reference configuration $\mathcal{B}_0$ with respect to the subsequent deformation through $\boldsymbol{F}$ hence contains a frozen-in anisotropic molecular orientation.

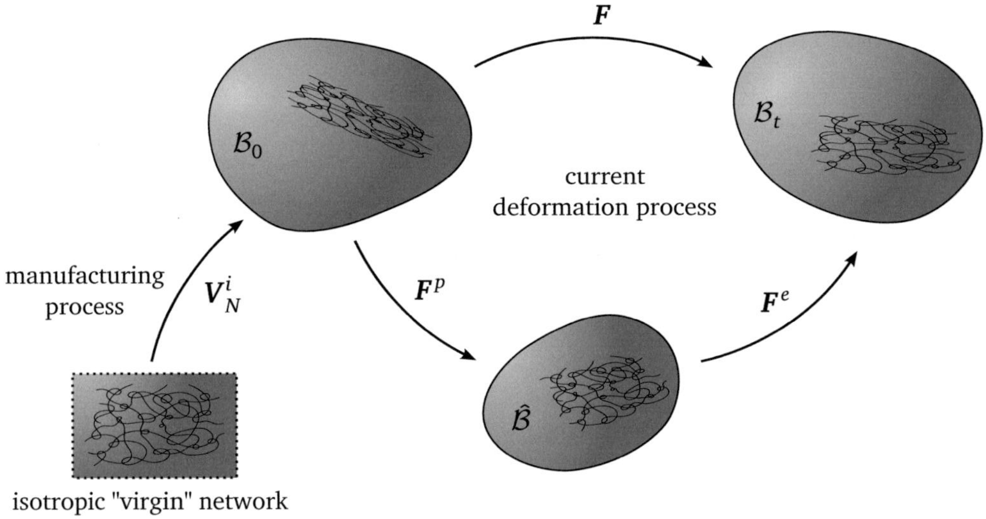

Figure 4.13.: Effect of initially anisotropic molecular network due to a preceding manufacturing process

The backstress is taken to be a function of $\boldsymbol{F}^p_S$ through the network state left Cauchy-Green tensor

$$\hat{\boldsymbol{b}}^p_S = \boldsymbol{F}^p_S \boldsymbol{F}^{p^T}_S \quad . \tag{4.65}$$

Since the kinematics of the "basic" and "entanglement dissociation" models are different to that of the "molecular relaxation by reptation" model the backstress depends on different kinematic tensors. In case of the "basic" model (Sec. 4.1.2) and its extension by the "entanglement dissociation" model (Sec. 4.2.1) the deviatoric backstress tensor is written as

$$\hat{\boldsymbol{\tau}}^b = \frac{C^R \lambda_L \beta}{3\lambda_C} \hat{\boldsymbol{b}}^{p'}_S \tag{4.66}$$

where the mean chain stretch is given by $\lambda_C = \left(\frac{1}{3}\mathrm{tr}[\hat{\boldsymbol{b}}^p_S]\right)^{\frac{1}{2}}$. In case of the "molecular relaxation" model (Sec. 4.2.2) the backstress depends on the network state left Cauchy-Green tensor obtained from

$$\boldsymbol{F}^e_{NS} = \boldsymbol{F}^e_N \boldsymbol{V}^i_N \quad , \quad \hat{\boldsymbol{b}}^e_{NS} = \boldsymbol{F}^e_{NS} \boldsymbol{F}^{eT}_{NS} \tag{4.67}$$

with the mean stretch $\lambda_C = \left(\frac{1}{3}\mathrm{tr}[\hat{\boldsymbol{b}}^e_{NS}]\right)^{\frac{1}{2}}$. For the total network stretch λ^L_N used in (4.60) the initial network stretch must be considered as well since it occurs in the left inelastic Cauchy-Green tensor

$$\hat{\boldsymbol{b}}^p = \boldsymbol{F}^p \boldsymbol{F}^{pT} = \boldsymbol{F}^e_N \boldsymbol{V}^i_N \boldsymbol{F}^p_N (\boldsymbol{F}^e_N \boldsymbol{V}^i_N \boldsymbol{F}^p_N)^T = \boldsymbol{F}^e_{NS} \tilde{\boldsymbol{b}}^p_N \boldsymbol{F}^{eT}_{NS} \quad . \tag{4.68}$$

4.3.2. Model response

In this section the response of the model is evaluated for different amounts of the initial stretch at room temperature and under uniaxial tension. In Tab. 4.6 the applied initial stretch values are summarized along with the corresponding (initial) mean chain stretch λ_C.

Table 4.6.: Max. applied pre-stretch λ^i_{N1}, corresponding stretch in perpendicular directions ($\lambda^i_{N2} = \lambda^i_{N3}$) and mean chain stretch λ_C. The limit chain stretch of the material is again set to $\lambda_L = \sqrt{2} = 1.414$

λ^i_{N1}	$\lambda^i_{N2} = \lambda^i_{N3}$	λ_C
1	1	1
1.2	0.9129	1.0176
1.4	0.8452	1.0628
1.6	0.7906	1.1269
1.8	0.7454	1.2043
2	0.7071	1.2910

In Fig. 4.14 the true stress-log. strain response of the model in the direction of the max. prescribed pre-stretch (max. eigenvalue $\lambda^i_{N1} \geq 1$) and perpendicular to that direction ($\lambda^i_{N2} = \lambda^i_{N3} \leq 1$) is shown. The response is qualitatively in agreement with the experimental data by Arruda et al. (1993) shown in Fig. 3.8. In the direction of the applied max. pre-stretch, the yield stress and the subsequent hardening strongly increase with increasing initial stretch. Perpendicular to that direction the opposite trend is observed: the higher the initial stretch the lower are the yield stress and the hardening. The extended material model is thus able to reproduce the material behavior in presence of a molecular pre-stretch, at least qualitatively.

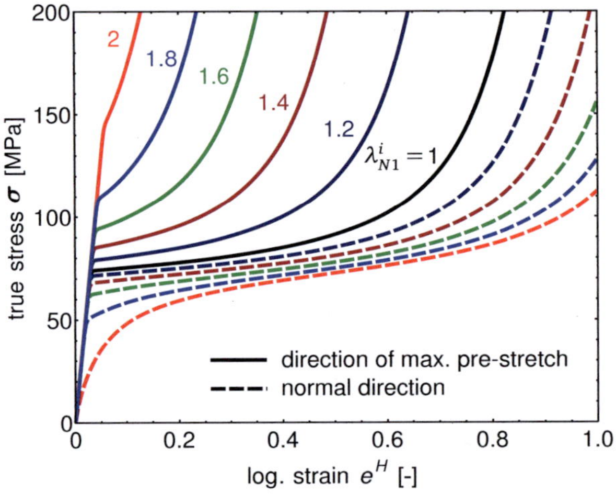

Figure 4.14.: Comparison of model response at different amounts of pre-stretch in direction and normal direction of max. pre-stretch ($\lambda^i_{N1} \geq 1$)

4.3.3. Simulation of pre-stretch induced dimensional instabilities

Manufacturing processes on amorphous thermoplastic polymers at elevated temperatures (e.g. injection molding) and subsequent rapid cooling may result in an anisotropic microstructure due to "frozen-in" molecular orientation (cf. Ch. 1). This state of pre-orientation in the solid material may give rise to stress-free deformations of a component during re-heating (Struik, 1990). This "memory-effect" can be observed at the injection molded plate shown in Fig. 4.15. After uniformly heating above the glass transition temperature the plate buckles due to the pre-oriented microstructure (Fig. 4.16a,b).

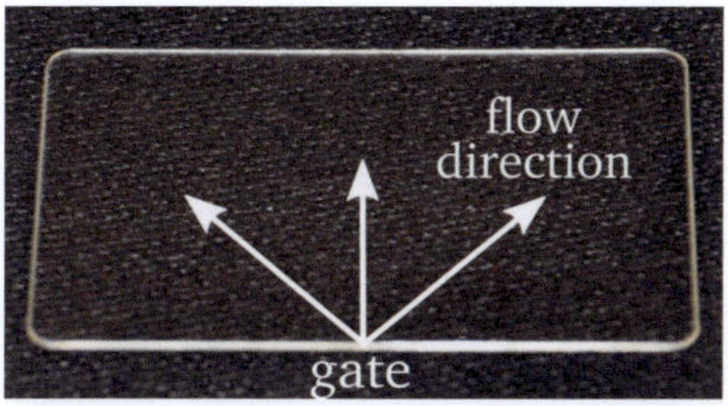

Figure 4.15.: Injection molded plate with assumed flow direction of melt

a) 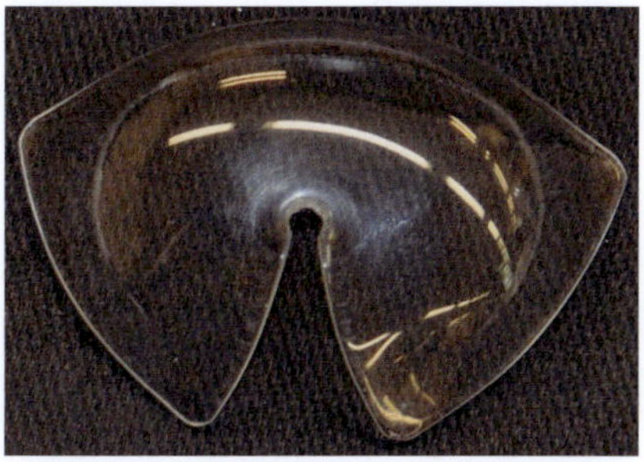

b) 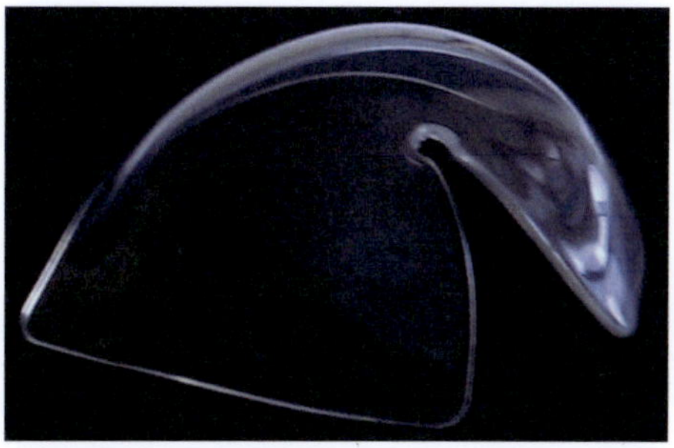

Figure 4.16.: Deformation of the plate after heating to 120 °C

As a computational example the injection molded plate is investigated. It is assumed that the highly stretched polymer melt flows radially into the mold and that the pre-stretch decreases linearly with increasing distance to the gate (filling point) according to a radial symmetric isochoric plane flow field (Fig. 4.15). A corresponding distribution and orientation of the maximum principal stretch λ^i_{N1} with five sections of different constant values of λ^i_{N1} (with $\lambda^i_{N2} = \lambda^i_{N3} = 1/\sqrt{\lambda^i_{N1}}$) is considered in the FE-model (Fig. 4.18a). The plate is subjected to a temperature which is taken spatially uniform and increases with time (Fig. 4.17). In addition, traction-free boundary conditions are imposed.

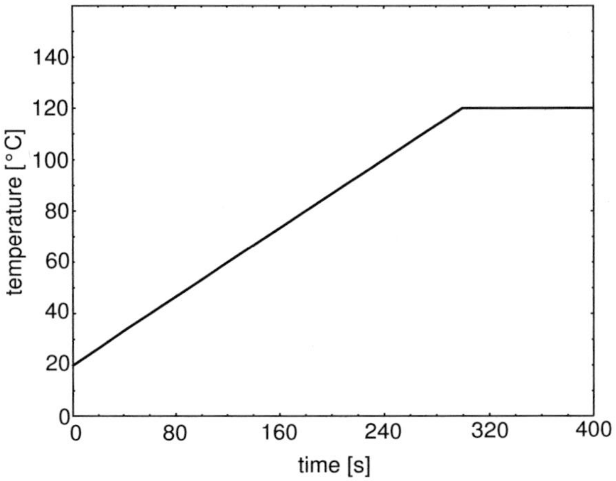

Figure 4.17.: Temperature variation over time

With increasing temperature the material flow resistance decreases and enables the "frozen-in" molecular stretch and backstress to relax. This spatially non-uniform re-deformation causes buckling of the plate. The deformed FE-model is presented in Fig. 4.18b,c which shows a qualitative good agreement with the deformed shape of the real component (Fig. 4.16a,b).

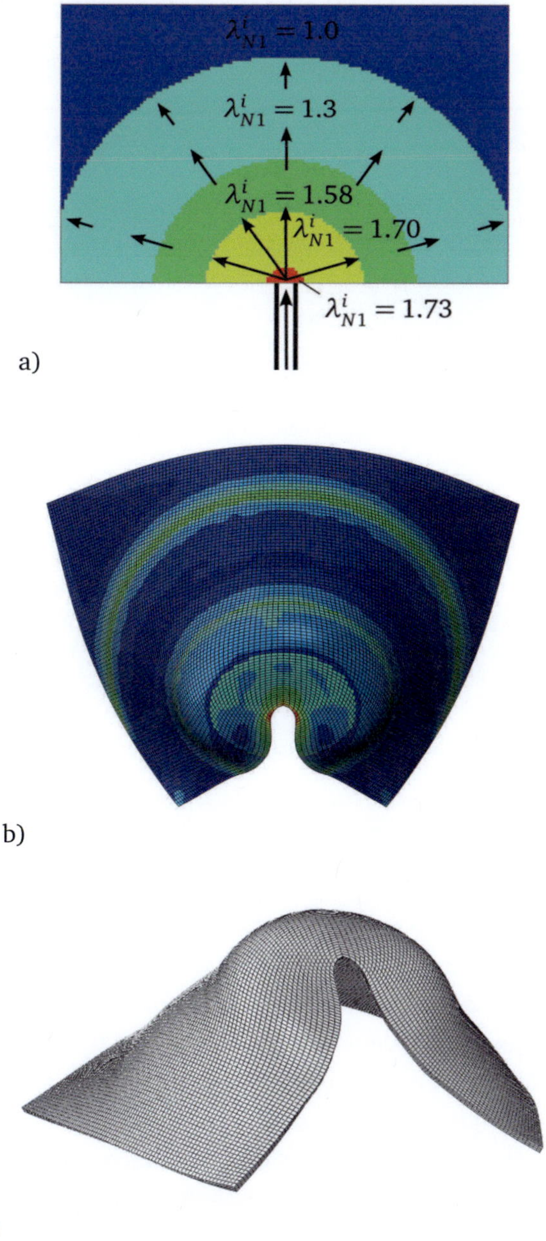

Figure 4.18.: FE analysis of the injection molded plate: a) orientation of initial molecular stretch and distribution of equivalent backstress $\tau_e^b = \sqrt{\hat{\boldsymbol{\tau}}^b : \hat{\boldsymbol{\tau}}^b}$ (max. value $\tau_e^b = 41.1$ MPa), b) distribution of equivalent backstress after annealing (max. value $\tau_e^b = 3.3$ MPa), c) deformation after heating to 120 °C

An injection molded tensile test specimen (Fig. 4.19a) serves as a second example. The deformed shape of the specimen after annealing is presented in Fig. 4.19b. The same assumptions as for the plate are applied for the computational model of the tensile test specimen. The orientation of the initial stretch and the corresponding equivalent backstress distribution are shown in Fig. 4.20a. The deformed FE-model after heating is presented in Fig. 4.20b.

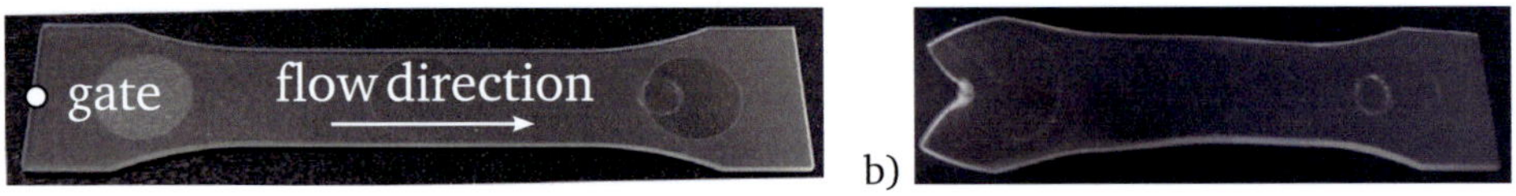

a) b)

Figure 4.19.: Injection molded tensile test specimen: a) initial state with indicated gate and flow direction, b) deformed state after heating to 120 °C

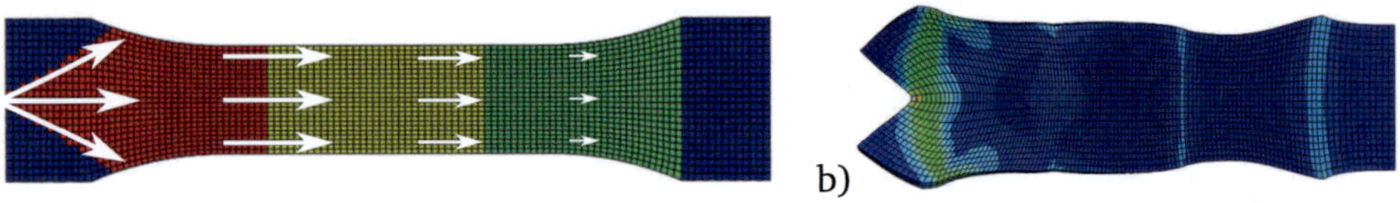

a) b)

Figure 4.20.: FE analysis of injection molded test specimen: a) orientation of initial molecular stretch and corresponding equivalent backstress distribution (max. value $\tau_e^b = 41.1$ MPa, same initial stretch values are used as for the plate), b) deformation to nearly stress free state after heating to 120 °C (max. value of equivalent backstress $\tau_e^b = 7.1$ MPa)

The constitutive model is used successfully to simulate a re-heating process of injection molded components. The resulting deformations of both parts are in good agreement with those observed in the real experiments. The correlation between the deformations of the real components and the deformations obtained in the simulations evidences that the assumed initial molecular orientation and its mapping to the computational models is reasonable. However, while the flow field in the present case is quite obvious, it is hardly conceivable that this heuristic method is applicable to more complicated geometries. In that case, data obtained from mold filling simulations might be used.

5. Simulation of thermoforming

In this chapter the simulation of the polymer film thermoforming process is investigated by applying the constitutive model developed in Sec. 4.2.2 and is implemented through a UMAT in the finite element software Abaqus. First, thermoforming of micro parts is simulated and compared to experimental results to validate the constitutive model. Two different computational models are employed and compared: a simplified 2D (plane strain) model and a full 3D model. As a second computational example, thermoforming of a blister like part is performed to investigate the influence of pre-streched films on the forming behavior. For this purpose, different amounts of pre-stretch in the film material are applied.

Thermoforming is a common method of processing thermoplastic polymers. The range of products made by thermoforming spans from simple packaging products to complicated parts, e.g. for the automotive or aircraft industry (Engelmann, 2012). In the general thermoforming process, polymer sheets or films are formed into a mold under applied pressure and temperature. In the first process step, the polymer is clamped between the mold and holding plates followed by the evacuation of the mold. The polymer is then heated close to the glass transition temperature T_g by thermal contact. In this temperature range, amorphous thermoplastic polymers are easy to form and large strains can be applied without any fracture, cf. Ch. 3. Pressurized gas is applied to form the polymer into the mold. After forming the mold is cooled and the part can be ejected. This process with the described four steps is depicted in Fig. 5.1.

The most relevant parameters in the thermoforming process affecting the forming and thus the quality of the final product are the time histories of the applied pressure and temperature. The mechanical behavior of the final product is mainly influenced by the film thickness and the polymer molecular network orientation after forming. The simulation of the whole process may show the correlation between process parameter variations and the quality of the formed part. The film thickness distribution throughout the part can be investigated as well as stresses and strains after forming.

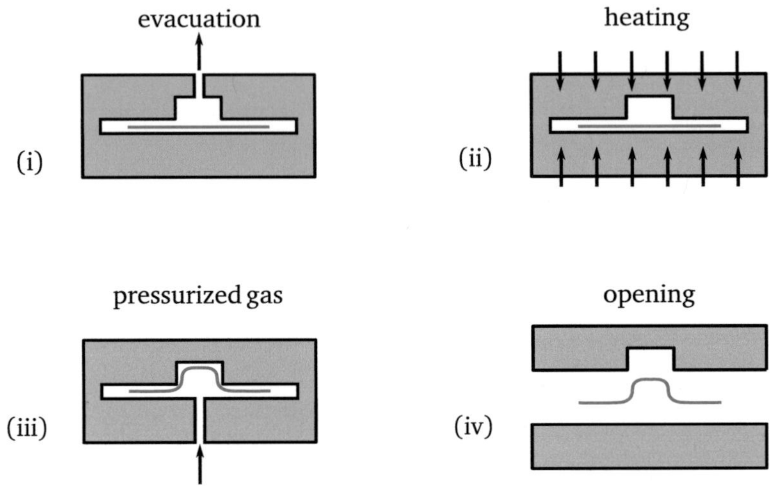

Figure 5.1.: Process steps of thermoforming: (i) A polymer film is clamped between the holding plate and the mold which is then evacuated. (ii) The tool is heated until the the glass transition temperature of the polymer film is reached. (iii) Pressurized gas forms the film into the mold. (iv) The film is cooled and the tool is opened.

Former works of simulating the thermoforming process are limited by either on considering the temperature dependent viscoplastic behavior of the material or to take all process steps into account. For example, Carlone and Palazzo (2006), McCool and Martin (2011), Kouba et al. (1992) and Nam et al. (2000) used hyperelastic models, Warby et al. (2003) a viscoplastic model, Karamanou et al. (2006) and Kim et al. (2009) viscoelastic models to simulate thermoforming. In some of these works, the material models are even independent on the temperature. In contrast, elastic-viscoplastic temperature dependent models are used by O'Connor et al. (2013) or Makradi et al. (2007). Only the latter developed a model which is similar to the model in the present work. These works were able to simulate the thermoforming process quite well, but all authors only considered the polymer at high temperatures and the inflation step, thus neglecting the cooling step. In contrast, aus der Wiesche (2004) considered the cooling step but used a small-strain viscoelastic model only. The consideration of the cooling step is expected to show an influence of the process parameters on the mechanical behavior of the final product as well and thus should be considered in the simulation. Thus, in this work the whole process as discussed above is investigated and simulated using the material model developed in Sec. 4.2.2 which is implemented in the finite element software Abaqus with a user material routine (UMAT).

In the next section, the process micro-thermoforming is simulated and the computational results are compared to experimental findings.

5.1. Micro-thermoforming

Micro-thermoforming is basically thermoforming at the microscale and was developed at the Institute of Microstructure Technology (IMT), Karlsruhe Institute of Technology (KIT), e.g. Heilig et al. (2010). It can be combined with nanoimprinting to get a pre-structured polymer film to produce film microchips which can be used for biological cell cultivation, for example. This requires the usage of very thin films (<100 μm) and special techniques for the thermoforming process. However, in this work only unstructured films are investigated.

Thermoforming experiments were performed at IMT, KIT using a PMMA film (Degalan G7E) with a thickness of 84 μm. The film was formed into a bottle like mold sketched in Fig. 5.2. The finished part is shown in Fig. 5.3.

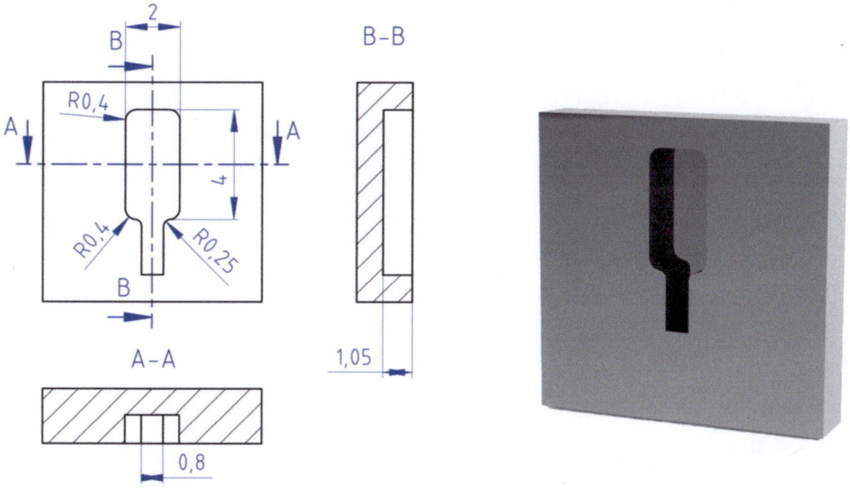

(a) 3D view of mold (b) dimensions of mold (in mm)

Figure 5.2.: The mold used in the experiments performed at IMT, KIT (Heilig, 2012)

The most important process steps and parameters of the experiments are summarized in the following, see also Tab. 5.1. First, the tool is heated with a temperature rate of 65°C/min to the final forming temperature. The film is then clamped with a force of 40 kN and the mold is evacuated (approx. 1 bar in 10 s). The forming pressure realized with pressurized gas is applied with a specific pressure increase rate and is held constant for a certain holding time afterwards. In the subsequent step, the tool is cooled with a cooling

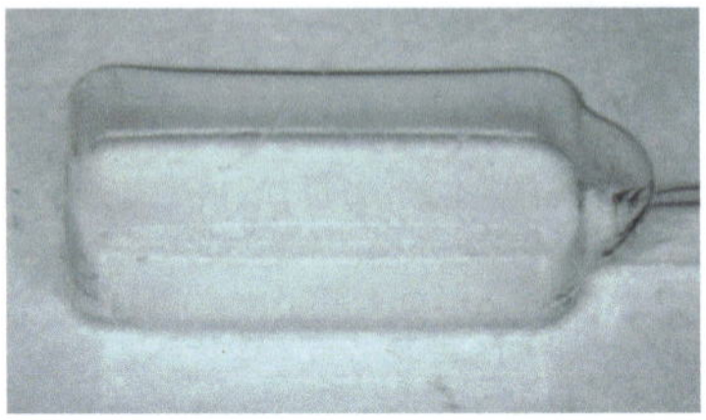

Figure 5.3.: The finished part after the thermoforming process

rate of 65°C/min till the demolding temperature is reached. The pressure is then released with a specific pressure reduction rate (Heilig et al., 2010). The process parameters of four representative experiments are given in Tab. 5.1. These four experiments (load cases) serve as examples for the validation of the simulation. The process parameters differ clearly and lead to a different forming behavior in the experiments.

Table 5.1.: Process parameters of four different experiments (Heilig, 2012)

load case #	forming temp. [°C]	pressure increase rate [MPa/s]	forming pressure [MPa]	pressure holding time [s]	pressure release temp. [°C]	pressure reduction rate [MPa/s]
1	110	0.3	2.5	55	75	0.3
2	115	0.4	4	100	90	0.4
3	105	0.4	1	100	90	0.4
4	105	0.4	1	10	90	0.2

The following simplifying assumptions are applied in all simulations. It is assumed that the deviation of the film temperature from the tool temperature is negligible during heating and cooling. Thus, a spatially constant temperature distribution is assumed. Rather, the same temperature is applied uniformly to the film which changes with time according to the temperature increasing/decreasing rate. Furthermore, the film is not clamped rather the nodes at the boundary of the film are fixed (Fig. 5.4).

For the tangential contact, the coefficient of friction $\mu = 0.3$ is assumed. Despite the film used in the thermoforming experiments and the material tested in Ch. 3 are both PMMA, it is likely that they slightly differ in the mechanical behavior. However, since there are no experimental data for the film, the results of Ch. 3 are used as input for the material model developed and calibrated in Sec. 4.2.2.

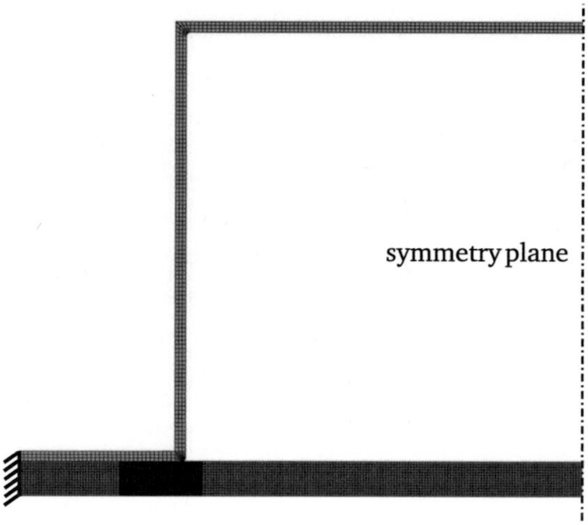

Figure 5.4.: 2D finite element model of the film (green) and the mold (gray)

Two different computational models are utilized in the following simulations. In a first approximation the simulations are done in 2D (plane strain) where the film in the center of the mold far away from the side walls is considered. In addition, the simulations are performed in 3D in which the complete mold is considered (Fig. 5.2). All simulations are conducted using the implicit static solver of Abaqus/Standard (Dassault Systèmes, 2012). The process steps in both computational models are listed in Tab. 5.2.

Table 5.2.: Process steps considered in the simulations

step	step description
1	evacuate mold (applying evacuation pressure)
2	increase pressure
3	hold pressure
4	decrease temperature
5	decrease pressure
6	open mold (release evacuation pressure)

5.1.1. 2D Simulation

The 2D plane strain model of the polymer film is meshed with ≈ 6500 reduced integration elements with mesh refinement in the region where the film comes into contact with the mold wall and the largest deformation is expected (Fig. 5.4). The mold is considered as rigid and, for imposing contact conditions, is discretized with rigid elements without any material behavior. The mesh is shown in Fig. 5.4.

The simulated film deformation after each step for the four different load cases is shown in Fig. 5.5 - Fig. 5.8. The major part of the deformation takes place in the first three steps (evacuation, increase and hold pressure) at all load cases. The last three steps (decrease of pressure and temperature, evacuation) have only minor influence on the final shape of the formed film. In load case #2, the highest forming temperature and pressure and a long pressure holding time were applied. This leads to the largest deformation of the film into the mold (Fig. 5.6f) in comparison to the other three simulated load cases. In load case #1, the residual forming depth is insignificantly lower (Fig. 5.5f). In this load case, a slightly lower forming temperature and a distinctly lower forming pressure along with a shorter holding time compared to load case #2 were applied. The process parameters of load case #3 and load case #4 only differ in the pressure holding time and the pressure reduction time but both have a much lower forming pressure and temperature than the load cases #1 and #2. The resulting forming depth of load case #4 compared to load case #3 is significantly lower (Fig. 5.7f and Fig. 5.8f). In summary, the higher the temperature, forming pressure and the longer the pressure holding time the more is the film formed into the mold. Since most of the deformation takes place in the first three steps, it is reasonable to consider only these when the thickness distribution in the part is of interest, as done in most of the earlier numerical studies mentioned above. However, if the stress and strain field after forming is of interest - for example for subsequent structural simulations - the whole process including the final steps must be considered.

The forming depth strongly correlates with the amount of max. principal strain which is shown for each load case after the last step in Fig. 5.9. The larger the forming depth the more the polymer is stretched and the more thinning of the polymer film occurs. For example, in load case #2, the minimum thickness of the film after the last step is half the initial thickness. In contrast, the thinning is about 17.5% only in load case #4. The max. local thinning simulated in each numerical load case is given in Tab. 5.3. The location of the max. thinning is for all load cases in the region of the mold edge and peak strains, respectively. The highly deformed elements in that region are shown for load case #2, for example.

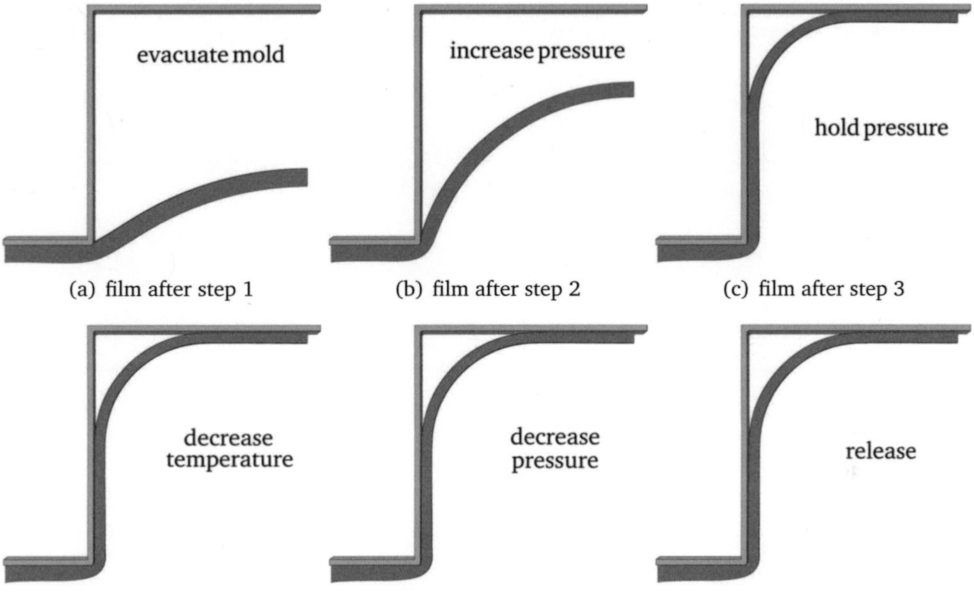

(a) film after step 1 (b) film after step 2 (c) film after step 3

(d) film after step 4 (e) film after step 5 (f) film after step 6

Figure 5.5.: Deformation of the film after each step of load case #1

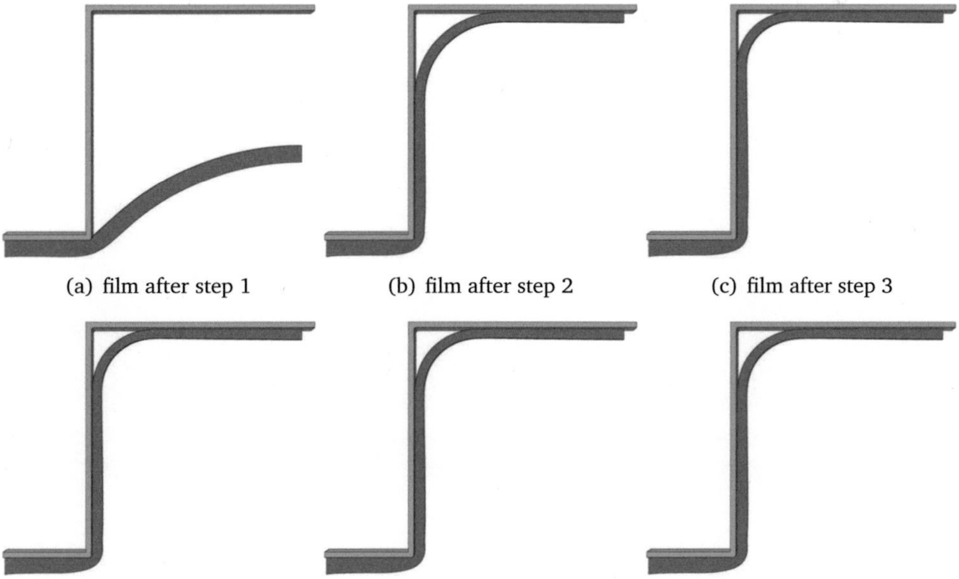

(a) film after step 1 (b) film after step 2 (c) film after step 3

(d) film after step 4 (e) film after step 5 (f) film after step 6

Figure 5.6.: Deformation of the film after each step of load case #2

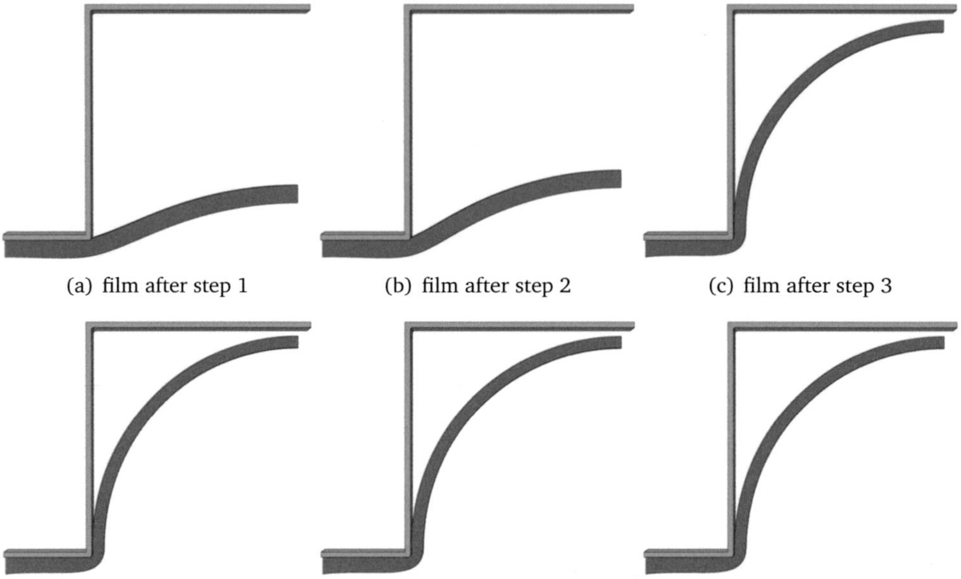

(a) film after step 1 (b) film after step 2 (c) film after step 3

(d) film after step 4 (e) film after step 5 (f) film after step 6

Figure 5.7.: Deformation of the film after each step of load case #3

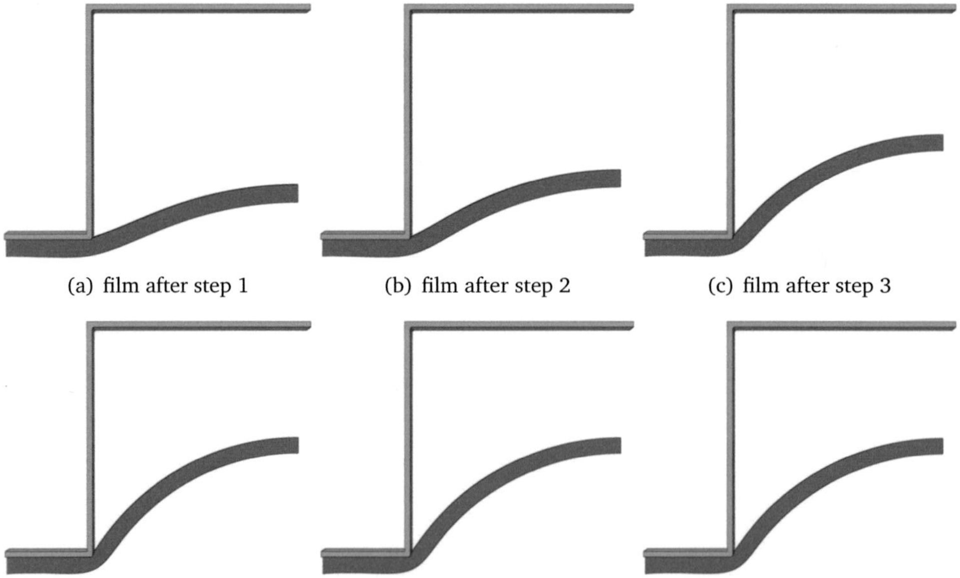

(a) film after step 1 (b) film after step 2 (c) film after step 3

(d) film after step 4 (e) film after step 5 (f) film after step 6

Figure 5.8.: Deformation of the film after each step of load case #4

Table 5.3.: Thinning of film (initial thickness 84 μm)

load case #	min film thickness [μm]	percentage change [%]
1	45.5	45.83
2	41.9	50.11
3	49.0	41.66
4	60.4	28.07

A comparison of the numerical results with real experiments by (Heilig, 2012) for the four load cases is shown in Fig. 5.10 in terms of the final shapes of the films. The forming depth is difficult to measure (Heilig, 2012) and the data obtained at the boundary of the mold are not reliable. Thus a meaningful quantity to compare the real experiments and the simulations is the maximum forming depth in the center of the mold summarized in Tab.5.4. For the first three load cases the max. forming depths are in very good agreement. Only in case #4 the deviation is relatively high. Despite the complex applied thermomechanical load, these results show that the computational model well captures the influence of individual process parameters on the forming behavior. Hence, the developed material model of Sec. 4.2.2 is obviously suitable for process simulations of amorphous thermoplastic polymers.

Table 5.4.: Comparison of max. forming depth

load case #	simulation [mm]	experiment [mm]
1	1.05	1.02
2	1.05	1.05
3	1.01	0.9
4	0.54	0.7

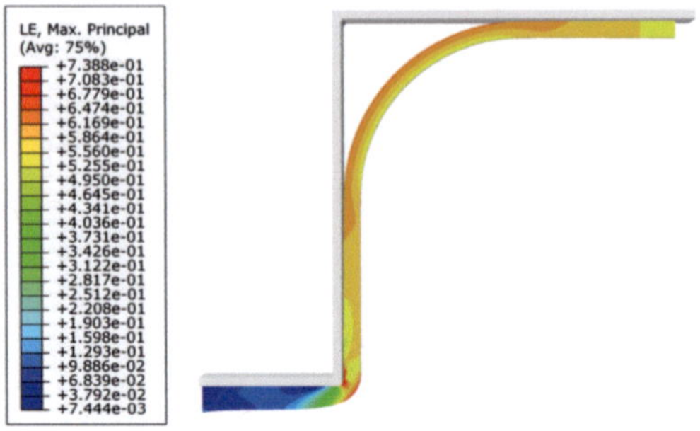

(a) loading case # 1

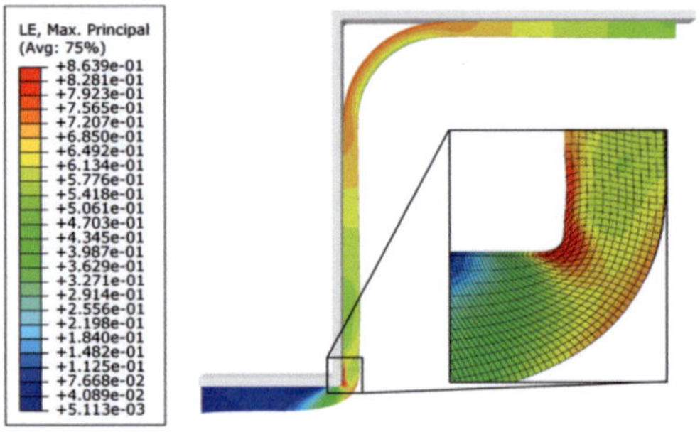

(b) loading case # 2

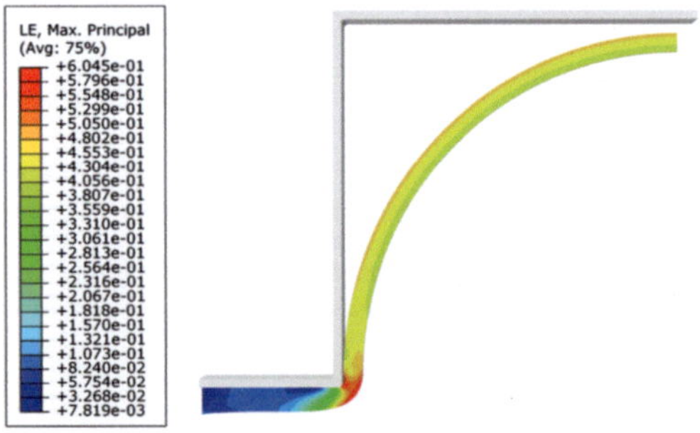

(c) loading case # 3

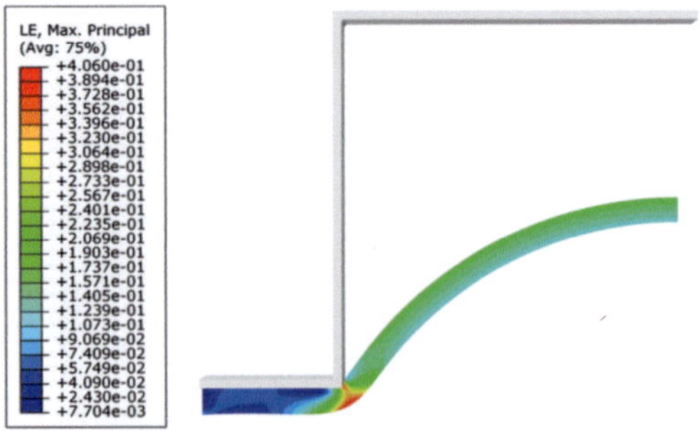

(d) loading case # 4

Figure 5.9.: Contour plots of max. principal strain after last step

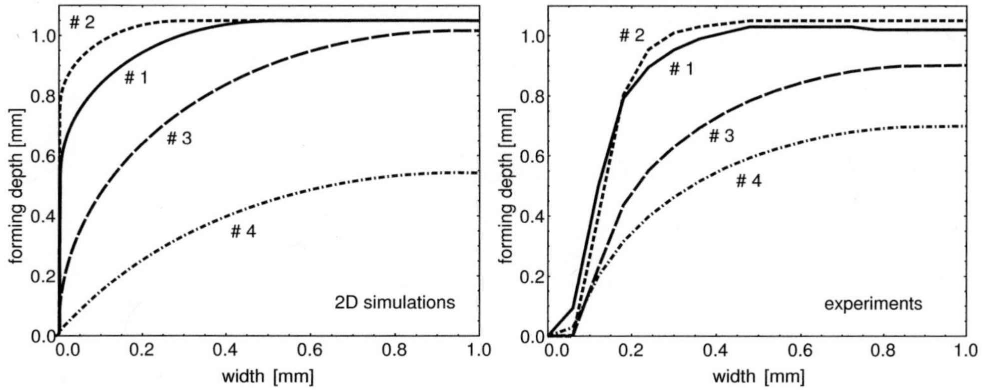

Figure 5.10.: Comparison of the resulting film shape in the numerical simulations (left) and the real experiments (right) measured with a tactile measurement device by Heilig (2012)

To investigate the influence of the coefficient of friction (between film and mold) this parameter is varied in simulations of load case #4 in which the largest deviation to the experiment occurred. Three different coefficients of friction ($\mu = 0$, $\mu = 0.1$, $\mu = 0.2$) are applied in addition to the originally one used ($\mu = 0.3$). In Fig. 5.11 the influence of friction on the simulated forming depth is shown. For higher values of the friction coefficient lower resulting forming depths are obtained. However, the influence of the coefficient of friction on the forming depth is relatively small in the range considered here. In reality, the coefficient of friction might be higher or rather the film sticks at the mold wall.

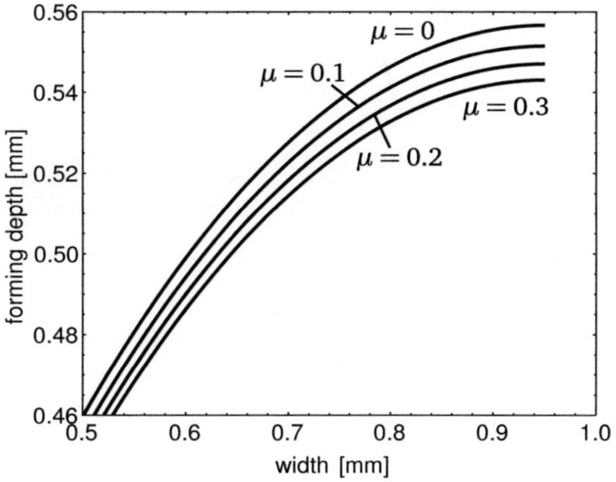

Figure 5.11.: Influence of friction in the simulation of load case #4

5.1.2. 3D Simulation

To investigate the influence of the plane strain simplification made in the above 2D simulations the whole mold (Fig. 5.2) is considered in the following. Symmetry of the problem is exploited and approximately 150000 reduced integrated solid elements are used to discretize the film. The same contact and friction behavior is applied as in the 2D model. The 3D model and the FE-mesh are shown in Fig. 5.12.

(a) 3D model of film is initially placed on the mold

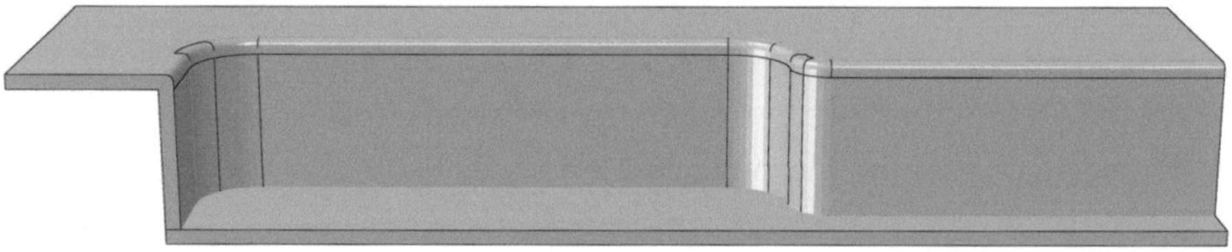

(b) 3D model of mold

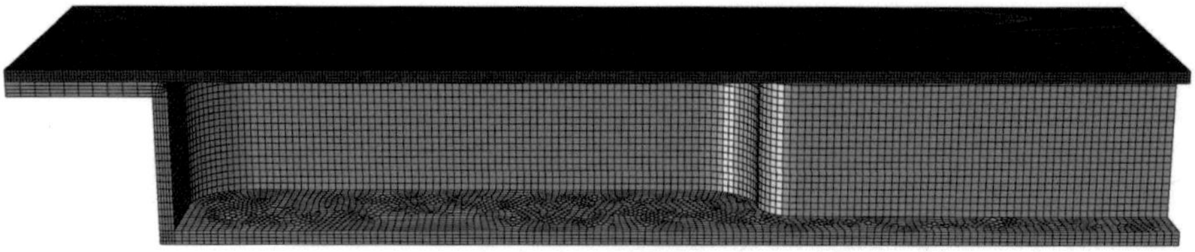

(c) FE mesh of the assembled model

Figure 5.12.: Half of 3D model and FE mesh of film and mold

The film deformation simulated with the 3D model is shown in Fig. 5.13 for each load case. In the comparison of the simulations to the experiments shown in Fig. 5.14 a good qualitative agreement is observable. In the center of the mold, a similar deformation behavior as in the 2D model is observed. The simulation of load case #2 again shows the deepest forming depth while load case #4 exhibits the lowest one.

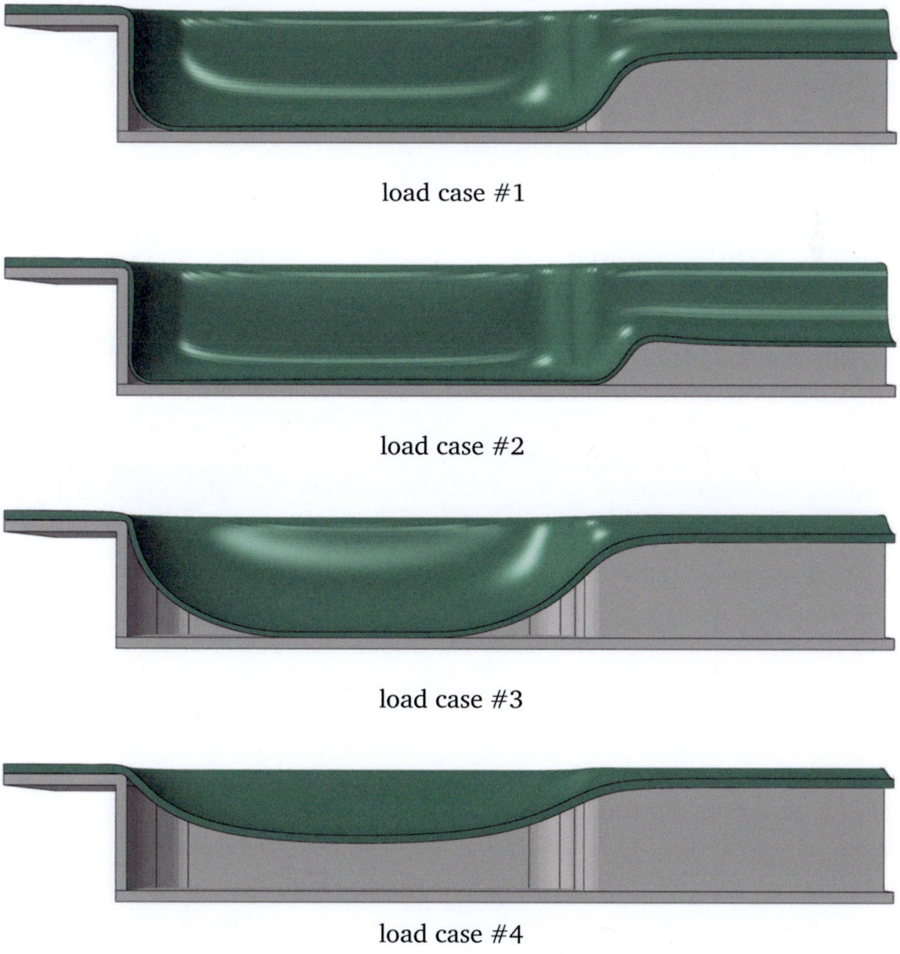

Figure 5.13.: Forming of the film into the mold

load case #1

load case #2

load case #3

load case #4

Figure 5.14.: Comparison of the forming behavior of the film in the simulations and the experiments

Max. principal strains are shown in Fig. 5.15 for each simulated load case. The largest values in the center of the parallel part of the mold differ only slightly from those of the 2D simulations (Fig. 5.9). However, the 3D simulations reveal significant differences in the region where max. strains are found in the final part for the different load cases (Fig. 5.15a and Fig. 5.15b). This cannot be reproduced with the 2D model of course.

(a) load case # 1

(b) load case # 2

(c) load case # 3

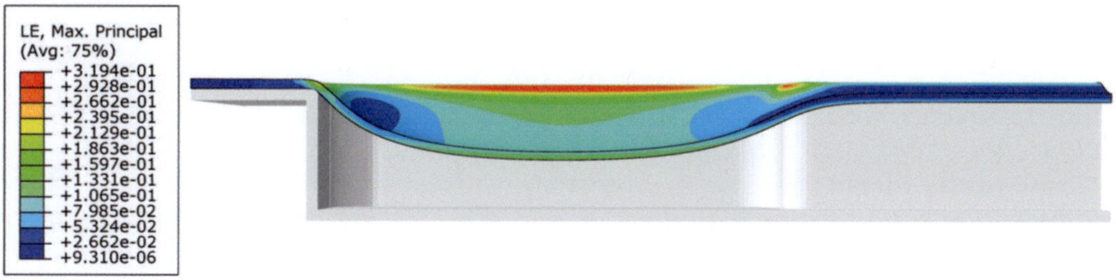

(d) load case # 4

Figure 5.15.: Contour plots of the max. principal log. strain after the last step for the four load cases

In Fig. 5.16 the shape of the formed film in the center of the mold is shown for the 3D simulations as well as for the 2D simulations for comparison reasons. It can be observed that the forming depth is slightly higher in the 3D simulations in all cases. This might be due to the less constrained film in the center region compared to the 2D simulations. However, the 2D model is suitable if the forming behavior in the middle of the mold is of interest only. Also it provides fast results since it consists of ≈ 20 times fewer degrees of freedom compared to the the 3D model.

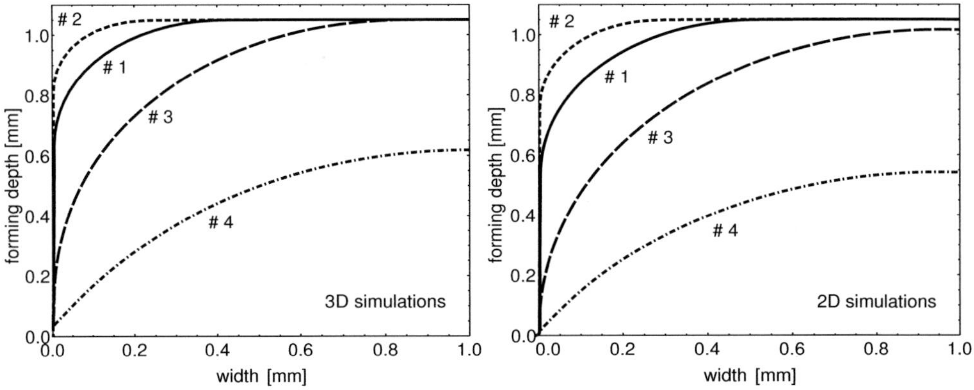

Figure 5.16.: Comparison of the forming depth of the 3D (left) and 2D (right) simulations. The forming depth of the 3D simulations is evaluated in the middle of the parallel part of the mold.

The results of the 2D and the 3D simulations showed a somewhat too large deformation of the film compared to the deformation in the experiment. This discrepancy may result from not clamping the film at the mold boundary so that the material is able to "flow" around the corner. Furthermore, the temperature may be not distributed uniformly over the whole film as assumed in the simulations which could lead to a lower deformation. In addition, it could be that a pre-stretched (and hence less deformable) film was used in the experiments which is not considered in the simulations. The influence of pre-stretch on the film deformation is investigated in the next section.

5.2. Influence of pre-stretch on thermoforming behavior

In this section, the influence of a pre-stretch on the forming behavior of polymer films is investigated. For this purpose an axi-symmetric mold ($r = 5$ mm) with a depth of 4.9 mm is considered in order to mimic the forming process of blisters which are used for packaging of drugs (Fig. 5.17). The film with a thickness of 0.5 mm initially lies flat on the top of the cavity. Because of the anisotropy of the pre-stretched film material the computational model can not be taken axi-symmetric. One-fourth (90° of the mold and the film is considered and meshed with $\approx$ 45000 linear hexahedral reduced integration elements for the film and $\approx$ 2000 rigid elements for the mold, respectively (Fig. 5.17).

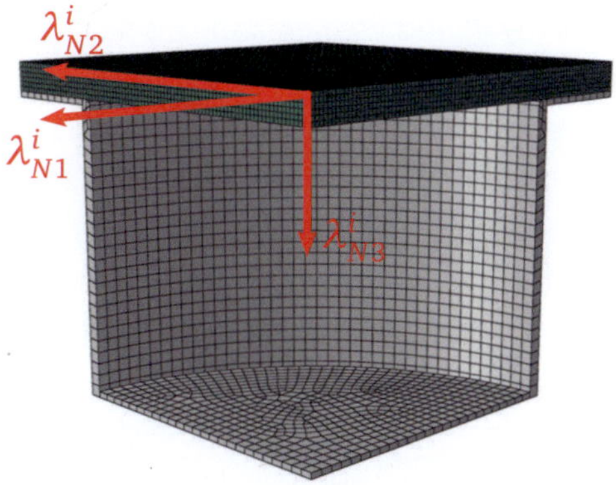

Figure 5.17.: FEM mesh of the film (green) and the mold (gray) giving a blister like form after thermoforming. The directions of applied eigenvalues of initial stretch tensor are indicated with red arrows.

The pre-stretch of the film is accounted for by prescribing the eigenvalues of the initial network stretch tensor in the constitutive model (see Sec. 4.3). The max. eigenvalue λ^i_{N1} is applied in the direction of one of the in-plane directions of the film as sketched in Fig. 5.17. The two other eigenvalues are taken to be equal with the requirement $\lambda^i_{N1}\lambda^i_{N2}\lambda^i_{N3} = 1$. The values of the applied eigenvalues are given in Tab. 5.5 for six different amounts of pre-stretch in the film material.

Table 5.5.: Six different initial stretches considered

λ^i_{N1}	$\lambda^i_{N2} = \lambda^i_{N3} = 1/\sqrt{\lambda^i_{N1}}$
1	1
1.2	0.912
1.4	0.845
1.6	0.790
1.8	0.745
2.0	0.707

Between the film and the mold a contact formulation is defined in the computational model. The simulation is divided into five steps listed in Tab. 5.6 representing a thermoforming process similar to that of the previous section but with fictitious process parameters. In the first step, the film (with spatially uniform temperature) is heated from room temperature up to the forming temperature over a certain period of time. This is necessary here to prevent a spontaneous relaxation of the backstress associated with the pre-stretch. The pressure is increased at constant forming temperature in the second step and in the third step the pressure is hold constant for one second. The temperature is decreased at constant pressure in the forth step and in the fifth step the pressure is released again at constant temperature. The process parameters are given in Tab. 5.6.

Table 5.6.: Process parameters

step	duration [s]	temperature [°C]	pressure [MPa]
1	1	20 - 110	0
2	1	110	0 - 2
3	1	110	2
4	1	110 - 20	2
5	1	20	2 - 0

The simulation for an unstretched film serves as a reference solution. The deformation behavior of this film is shown in Fig. 5.18 for all five steps. The largest deformation occurs during step two (pressure increasing) and step three (pressure holding).

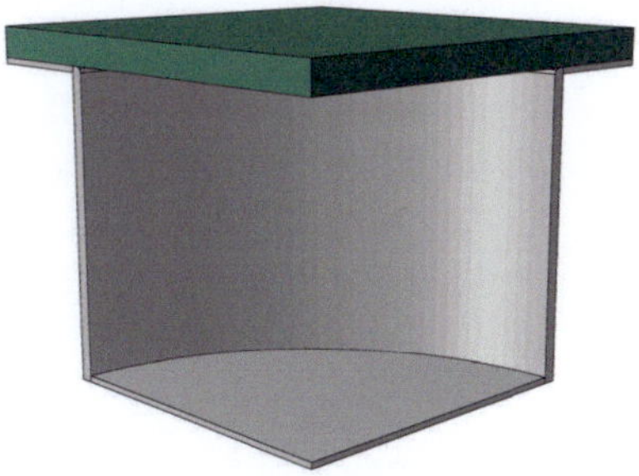

(a) film after step 1

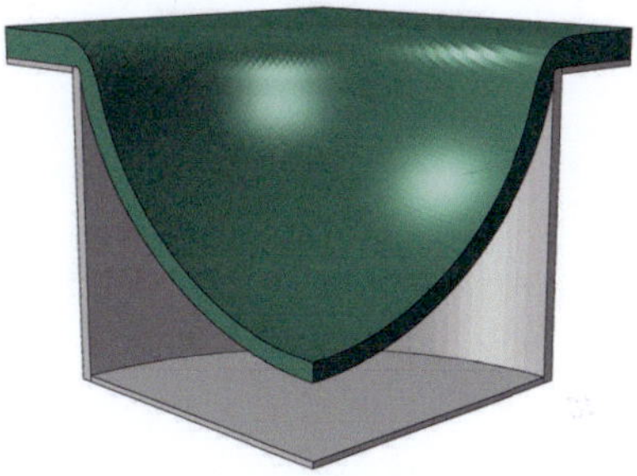

(b) film after step 2

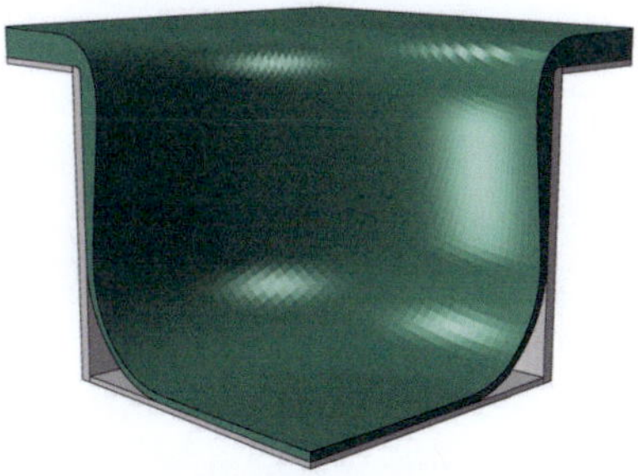

(c) film after step 3

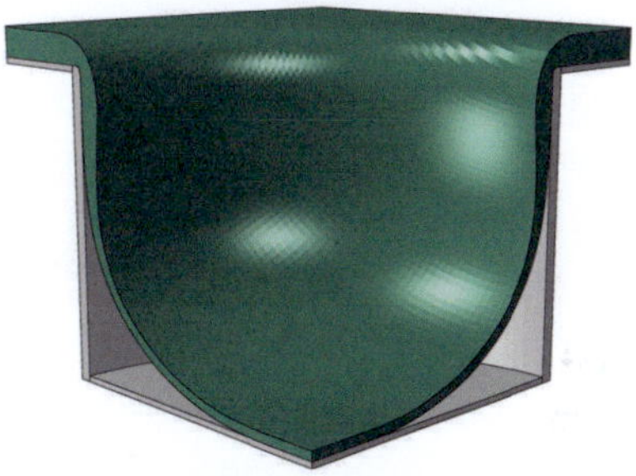

(d) film after step 4

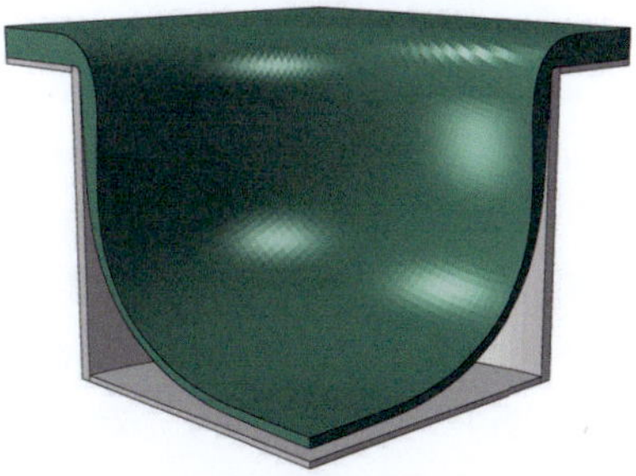

(e) film after film after step 5

(f) film after film after step 5

Figure 5.18.: Deformation of the film without initial pre-stretch after each step

In Fig. 5.19, the deformation of the differently pre-stretched films after the last step is shown. It is clearly observable that the forming depth is strongly dependent on the amount of pre-stretch: the more the film is pre-stretched the lower is the forming depth and the lower the local peak value of the max. principal strain (Fig. 5.20). Thus, the thinning of the film is dependent on the pre-stretch as well. The thinning is given in Tab. 5.7 for the differently pre-stretched films.

Table 5.7.: Thinning of each film after last step

film	minimum film thickness [mm]	percentage change [%]
$\lambda_{N1}^{i} = 1$	0.156	68.8
$\lambda_{N1}^{i} = 1.2$	0.177	64.6
$\lambda_{N1}^{i} = 1.4$	0.204	59.2
$\lambda_{N1}^{i} = 1.6$	0.237	52.6
$\lambda_{N1}^{i} = 1.8$	0.289	42.2
$\lambda_{N1}^{i} = 2.0$	0.336	32.8

Since the max. pre-stretch is applied in one of the in-plane directions of the film (see Fig. 5.17) its deformation behavior is anisotropic. Since the film is not clamped but evenly fixed at the outer boundary, the film is able to detach from the mold. This can be seen in Fig. 5.19e and Fig. 5.19f for large values of the pre-stretch. In Fig. 5.19f, the detach point is indicated by "A" while the film at point "B" is in contact with the mold wall. Due to the anisotropic material behavior a non-uniformity can be observed in the spatial distribution of the max. principal strain as well (Fig. 5.20). The max. strains occur at point "A" which is a direct result of the specified amount and direction of the max. principal pre-stretch (Fig. 5.20f): the material yield strength in the direction of bending at point "A" is reduced while at point "B" it has its maximum[1]. Hence, the resistance against bending around the particular bending axis at point "A" is reduced and at point "B" increased, respectively. A more pronounced anisotropic deformation behavior can not be observed due to the high forming temperatures. However, the results of this investigation show a strong influence of the pre-stretch on the film deformation behavior which should not be neglected in computational analyses of thermoforming.

[1] The influence of the pre-stretch on the mechanical behavior is discussed in Sec. 3.1.2 and Sec. 4.3

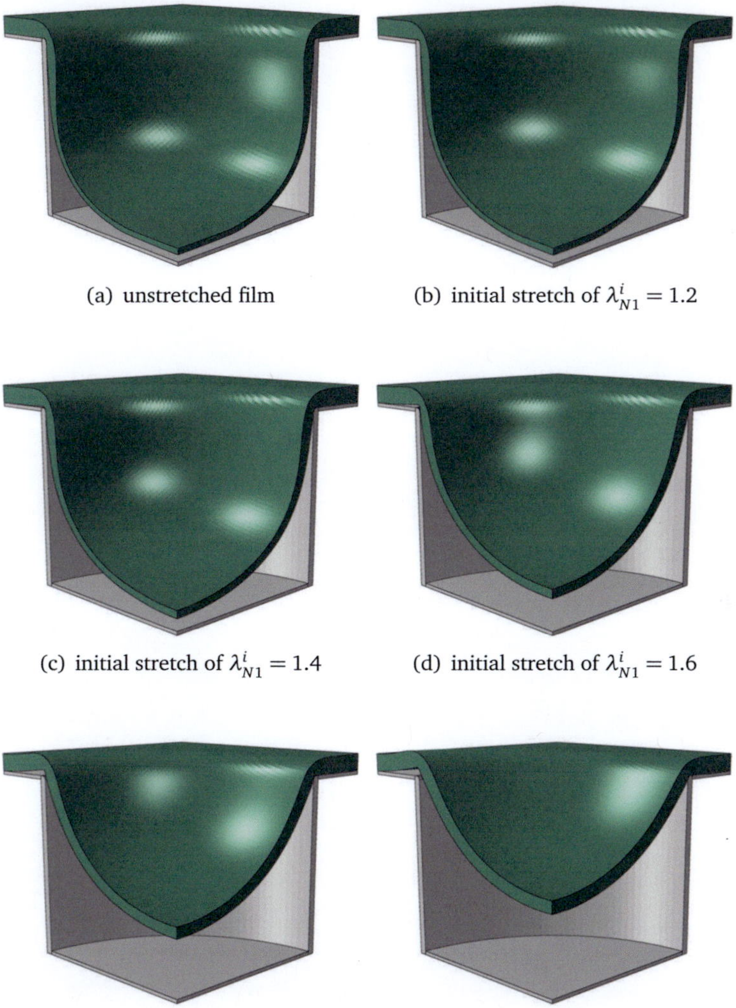

(a) unstretched film

(b) initial stretch of $\lambda^i_{N1} = 1.2$

(c) initial stretch of $\lambda^i_{N1} = 1.4$

(d) initial stretch of $\lambda^i_{N1} = 1.6$

(e) initial stretch of $\lambda^i_{N1} = 1.8$

(f) initial stretch of $\lambda^i_{N1} = 2$

Figure 5.19.: Deformation of each film after last step

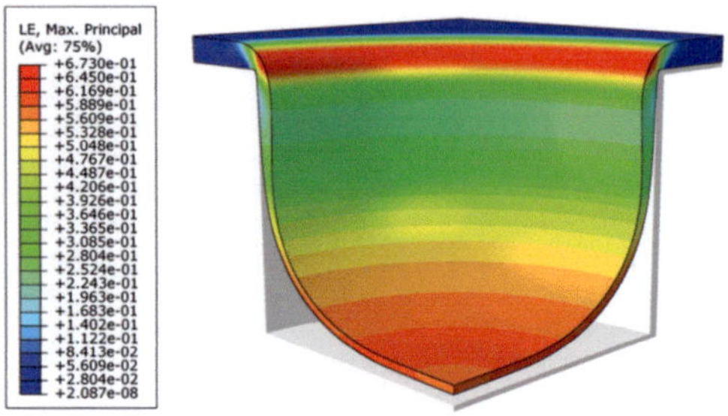

(a) unstretched film

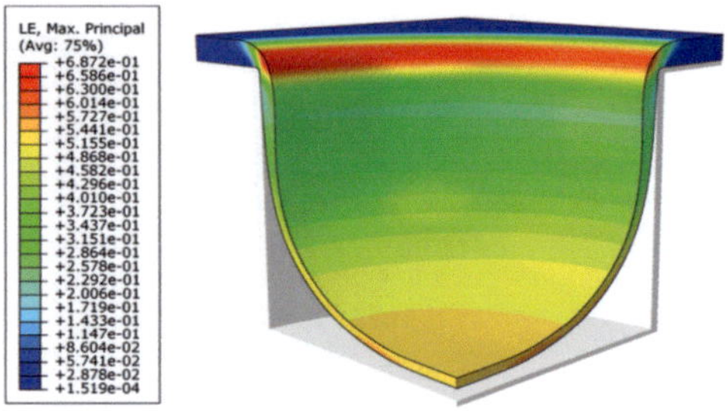

(b) initial stretch of $\lambda_{N1}^{i} = 1.2$

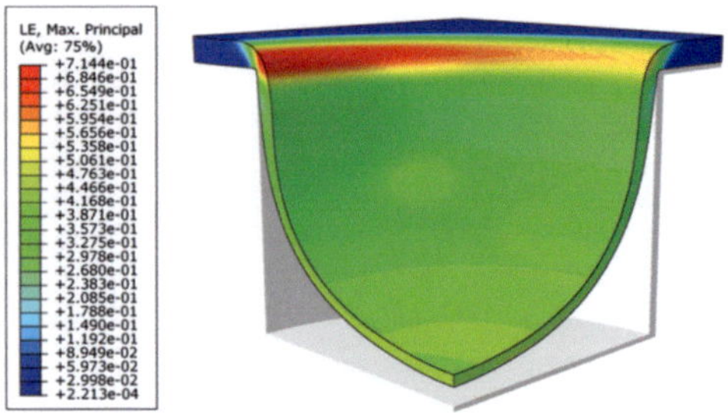

(c) initial stretch of $\lambda_{N1}^{i} = 1.4$

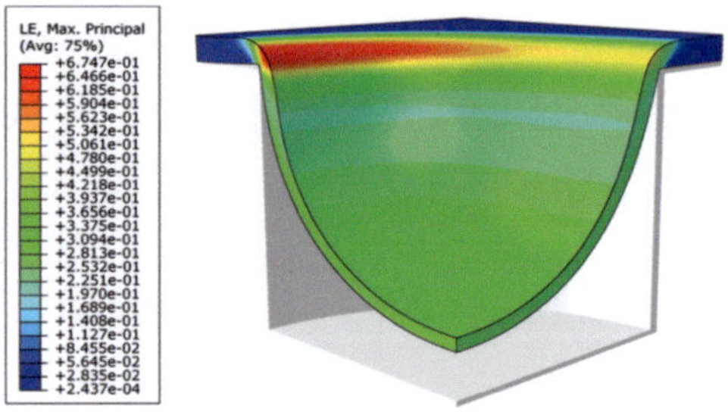

(d) initial stretch of $\lambda_{N1}^{i} = 1.6$

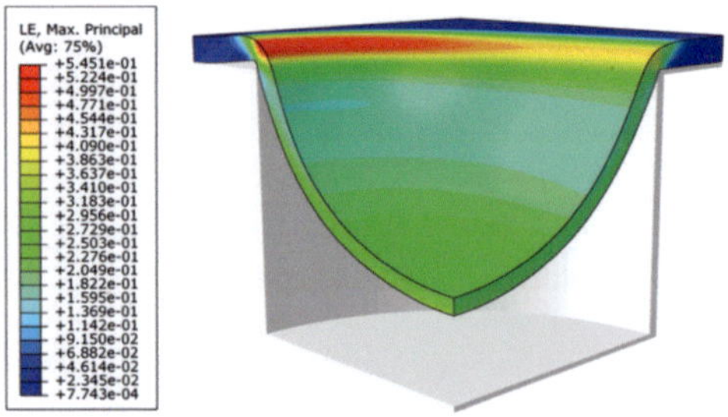

(e) initial stretch of $\lambda_{N1}^{i} = 1.8$

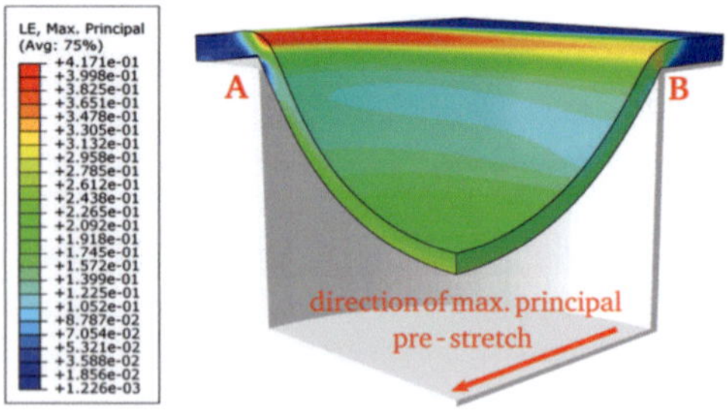

(f) initial stretch of $\lambda_{N1}^{i} = 2$

Figure 5.20.: Contour plot of max. principal log. strain in differently pre-stretched films after the last step

5.3. Conclusions from thermoforming simulations

From the above investigation of the simulation of thermoforming the following conclusions can be derived. The influence of various process parameters like temperature, pressure and holding time on the forming behavior of polymer films is well captured by the simulations. The numerical and the experimental results are found to be qualitative and quantitative in a reasonable agreement. This provides a good validation of the thermomechanical constitutive model developed and calibrated in Sec. 4.2.2.

However, to even improve the computational FE model more realistic boundary conditions may be applied. For example, one could consider the clamping of the film (apply a clamp force) or include a non-uniform temperature distribution by computing the heat transfer in the tool prior to the structural analysis or even perform a thermomechanical coupled simulation. Furthermore, a large coefficient of friction could have a strong influence on local deformation behavior and strain distribution of the film, e.g. sticking on the mold wall at high temperatures. Also, pre-stretched films showed a strong influence on the forming forming behavior and thus can not be neglected in thermoforming simulations.

6. Symbolic programming of user material routines

Some part of this work involves the implementation of the constitutive equations described in Sec.4 for the user material interface of the implicit solver Abaqus/Standard (UMAT). For this purpose the automatic differentiation, code generation and optimization tool AceGen (Korelc, 2012) is used to simplify the implementation process as well as to automatically form required derivatives. These may be the consistent tangent moduli, the Cauchy stress derived from a strain energy function for hyperelastic models or the Jacobian of a Newton-iteration scheme.

In this chapter, the application of AceGen will be described with the main functionalities[1] *used and how it helps in the material modeling and implementation process. Exemplarily, the implementation of a hyperelastic and an elastic-plastic material model utilizing AceGen will be presented. The hyperelastic model serves as an introductive example and is useful to illustrate the usage of AceGen. Furthermore, the computed derivatives can be verified with this model because they are relatively simple. The numerical treatment of the elastic-plastic material model is similar to that models of Ch. 4 but the constitutive equations are less complicated. Thus, it serves as an example where a local Newton-iteration scheme must be performed to update the internal variables. In this case, the consistent tangent moduli depends not only on the total deformation but also on the internal variables which leads to complicated derivatives. For both material models the stress and the consistent tangent moduli will be derived as needed in Abaqus/Standard. The consistent tangent will be verified by comparing it to the numerically approximated tangent which is also obtained using AceGen.*

The numerical treatment and implementation of the constitutive models of Ch. 4 using AceGen is given in App. A.2.

[1] The algorithms implemented in AceGen are not described in this work. Rather, AceGen is used as a "Black Box" and thus a verification is necessary.

6.1. Introduction to AceGen

The software tool AceGen is a plug-in for the software package Mathematica (Wolfram, 2013). It was developed by Prof. Dr.-Ing. J. Korelc (Korelc, 1996) to automatically differentiate equations and to generate and -at the same time- optimize program code, for example FORTRAN (Intel, 2013) code which may be used within the user routines of Abaqus. So far, AceGen was mainly used for the efficient implementation of finite elements, e.g. in Korelc and Wriggers (1999), Korelc (2002), Wriggers (2008), Mattern (2012). In this work, it is used to code constitutive equations and to apply needed algorithms. This brings big advantages for the implementation and material modeling process. These advantages are described in the following.

Since AceGen uses the symbolic programming environment of Mathematica, in which programming is fairly easy compared to a programming language like FORTRAN. The constitutive equations can be coded as they are "written on paper". An example: programming the right Cauchy-Green tensor (2.14) with FORTRAN is done by using loops and summing the respective products of the components of the deformation gradient or by coding every sum by hand. Even this small example might be time consuming and error-prone to implement. Using AceGen the product of the right Cauchy-Green tensor in Mathematica and the AceGen syntax is simply given by $\mathbb{C} \models \mathbb{F}^{\mathrm{T}} . \mathbb{F}$ where the symbol $\models$ is an AceGen specific operator for the symbolic field $\mathbb{C}$ and "." is the Mathematica command for a tensor product, respectively. So, besides using AceGen commands and functions which begin with `SMS` (Symbolic Mechanics System) it is possible to use the built-in Mathematica functions as well. The example above shows that this way of programming is neither time consuming nor error-prone. It highly minimizes programming errors and debugging is not necessary in most cases. A second advantage follows directly from this symbolic way of programming: equations are quickly exchangeable. So, it is possible to analyze different approaches without re-programming a whole new routine. For example, if one would test different flow functions the only thing to change is the equation for that function. The remaining algorithm would be untouched and no new derivatives need to be computed by hand. The automatic symbolic differentiation of equations is a further big advantage. Especially performing complicated derivatives by hand might be a time consuming task. While this functionality is available in many programs, the advantage of AceGen is the optimization of the generated code which is selectable along with the choices of the programming languages: C, C++, FORTRAN, Mathematica or Matlab. In this work, only FORTRAN code is used for the UMAT routines. An optimized code used for numerical simulations implies a faster computation due to less operations performed. Particularly, the efficiency of the code is important for an explicit FEM code where less operations are important to reduce computation times (Mattern, 2012). It is also important for implicit FEM code if a large number of degrees of freedom is used. A conceptual drawback of the code generation is the impossible interpretability of the code for a reader in sense of debugging because all operations are performed with auxiliary variables.

Summing up, the tool AceGen in combination with Mathematica accelerates programming and consequently saves time. It takes care of the time consuming aspects of the material model implementation by automatically computing derivatives and the flexibility of the symbolic programming. It automatically generates code which is, at best, more efficient than code programmed by hand in terms of the needed number of operations.

In the next two sections it will be shown how to use AceGen to generate a UMAT routine for specific material models and how to verify the result. The used AceGen commands and functions are briefly explained. For a detailed description it is referred to the manual (Korelc, 2009).

6.2. Implementation of material models using AceGen

In this section, the implementation of a hyperelastic and an elastic-plastic material model using AceGen for a UMAT routine in Abaqus/Standard is presented. The UMAT interface requires the calculation of the Cauchy stress and the tangent moduli. Note, only the Cauchy stress is needed for a correct solution if a solution is obtained. The latter depends on the computed tangent moduli which determines the convergence rate of the Newton-scheme for solving a mechanical boundary value problem with the finite element method. If the tangent moduli are computed with respect to the underlying integration algorithm a quadratic convergence of the Newton scheme is obtained (Hughes and Pister (1978), Simo and Taylor (1985), Simo and Hughes (1998)). This tangent is called "consistent" or "algorithmic" tangent because the underlying update algorithm is consistently linearized and differs from the continuum tangent.

The analytical derivation of the consistent tangent moduli may be a difficult task if the material model is complicated. To overcome this problem one could use either the feature of symbolic computation and automatic differentiation in a software package (e.g. Mathematica (Wolfram, 2013), Matlab (Mathworks, 2013), OpenAD (OpenAD, 2013)) or a numerical approximation method (Miehe (1996), Sagar and Stein (2008), Sun et al. (2008), Tanaka et al. (2014)). The algorithm needed for an approximation method might be very long but gives at best the same convergence rate as the analytical tangent. As stated before, in this work AceGen is used to compute the Cauchy stress and the consistent tangent.

6.2.1. Hyperelastic material model

The hyperelastic isotropic neo-Hookean material model (same as the built-in in Abaqus/-Standard) is chosen in this section to demonstrate and verify the implementation with AceGen. The tangent moduli are given in (Dassault Systèmes, 2012) and are programmed with AceGen as well to get an optimized code for comparison reasons. The computation with AceGen utilizing the automatic differentiation technique should give the same results as the solution derived "by hand". This is verified with the analytical tangent.

A hyperelastic material model is defined by a strain energy function which depends on the deformation, e.g. the right Cauchy-Green tensor $W = W^{\dagger}(\boldsymbol{C})$. By definition of a perfectly elastic material the dissipation is zero and the Cauchy stress follows directly from (2.94) for isothermal conditions as

$$\boldsymbol{\sigma} : \boldsymbol{d} = J^{-1}\dot{W} = J^{-1}\frac{\partial W}{\partial \boldsymbol{C}} : \dot{\boldsymbol{C}} = J^{-1}2\boldsymbol{F}\frac{\partial W}{\partial \boldsymbol{C}}\boldsymbol{F}^T : \boldsymbol{d} \tag{6.1}$$

$$\Rightarrow \boldsymbol{\sigma} = J^{-1}2\boldsymbol{F}\frac{\partial W}{\partial \boldsymbol{C}}\boldsymbol{F}^T = J^{-1}\boldsymbol{F}\boldsymbol{S}\boldsymbol{F}^T \tag{6.2}$$

with the second Piola-Kirchhoff tensor $\boldsymbol{S} = 2\frac{\partial W(\boldsymbol{C})}{\partial \boldsymbol{C}}$.

For the objective stress rate calculation in Abaqus/Standard the Jaumann-Zaremba rate of the Cauchy stress (2.54) is used (Dassault Systèmes, 2012). The rotational part of the stress update is done by Abaqus so that the user has to compute the stress update only. The tangent required by Abaqus for a UMAT routine is given by the linearization of the Kirchhoff stress $\boldsymbol{\tau}$ in terms of the Jaumann-Zaremba stress rate

$$\Delta\boldsymbol{\tau} = \tilde{\mathbb{C}} : \Delta\boldsymbol{d} + \Delta\boldsymbol{w}\,\boldsymbol{\tau} - \boldsymbol{\tau}\Delta\boldsymbol{w} \quad . \tag{6.3}$$

Using the relation defined in (2.110) the tangent modulus is found to be

$$\tilde{\mathbb{C}}_{ijkl} = \mathbb{C}_{ijkl} + \frac{1}{2}(\delta_{ik}\tau_{jl} + \tau_{ik}\delta_{jl} + \delta_{il}\tau_{jk} + \tau_{il}\delta_{jk}) \quad . \tag{6.4}$$

The tangent to implement in the UMAT routine must be related to the Cauchy stress and consequently $\tilde{\mathbb{c}}$ must be divided by $J = \det[\boldsymbol{F}]$

$$\tilde{\mathbb{c}}^{\sigma} = J^{-1}\tilde{\mathbb{c}} \quad . \tag{6.5}$$

The spatial elasticity tensor $\mathbb{c}$ in (6.4) is obtained by the push-forward of the material elasticity tensor (see (2.109)) which is the second derivative of the strain energy function with respect to the right Cauchy-Green tensor

$$\mathbb{C} = 2\frac{\partial \boldsymbol{S}}{\partial \boldsymbol{C}} = 4\frac{\partial^2 W(\boldsymbol{C})}{\partial \boldsymbol{C}\partial \boldsymbol{C}} \quad . \tag{6.6}$$

The neo-Hookean strain energy function used in the following reads

$$W(\boldsymbol{C}) = C_{10}(\check{I}_C - 3) + \frac{1}{D_1}(J-1)^2 \tag{6.7}$$

with the material parameters C_{10} and D_1 and the relations for the isochoric invariants (cf. (2.8))

$$\check{I}_C = \det[\boldsymbol{C}^{-1/3}]\,\mathrm{tr}[\boldsymbol{C}] = III_C^{-1/3} I_C \tag{6.8}$$

and

$$J = \det[\boldsymbol{F}] = III_C^{1/2} \quad . \tag{6.9}$$

The Cauchy stress follows from (6.2) and reads

$$\boldsymbol{\sigma} = \frac{2}{J}C_{10}(\check{\boldsymbol{b}} - \frac{1}{3}I_{\check{b}}\mathbf{1}) + \frac{2}{D_1}(J-1)\mathbf{1} \quad . \tag{6.10}$$

The computation of the tangent moduli is given in Box 6.1 and yields in index notation

$$\begin{aligned}
\tilde{\mathbb{c}}^{\sigma}_{ijkl} &= \frac{2}{J}C_{10}\Big(\frac{1}{2}(\delta_{ik}\check{b}_{jl} + \check{b}_{ik}\delta_{jl} + \delta_{il}\check{b}_{jk} + \check{b}_{il}\delta_{jk}) - \frac{2}{3}(\delta_{ij}\check{b}_{kl} + \check{b}_{ij}\delta_{kl}) \\
&+ \frac{2}{9}\delta_{ij}\delta_{kl}\check{b}_{mm}) + \frac{2}{D_1}(2J-1)\delta_{ij}\delta_{kl}\Big) \quad .
\end{aligned} \tag{6.11}$$

In the next sections, the algorithm to compute the stress and the analytical tangent with AceGen for a general hyperelastic material model is presented. As computational examples, a single element test and the simulation of a three point bending experiment of a rubber plate where large deformations occur are used to compare the different approaches. Within these examples the error with respect to the reference solution, the computation time, the number of iterations and the convergence rate are investigated.

Numerical treatment

The computation of the derivative (6.6) may be a time consuming task if the strain energy function is complicated (e.g. anisotropic) and is carried out by AceGen here. The algorithm to compute the stress and the consistent tangent is shown in Box 6.1. The items 1-6 are computed by AceGen and it is marked by (AD) where the automatic differentiation feature is used. The corresponding FORTRAN code is generated as a subroutine. This subroutine is called by the UMAT routine in which all quantities are declared and the stress and tangent are given back to the Abaqus main program.

Algorithm to compute the stress and the tangent moduli

input in UMAT: $\boldsymbol{F}_{t+\Delta t}$

$\rightarrow$ computation with AceGen :

1. right Cauchy-Green tensor: $\boldsymbol{C} = \boldsymbol{F}^T \boldsymbol{F}$
2. strain energy function: $W(\boldsymbol{C})$
3. stress tensors (AD): $\boldsymbol{\tau} = 2\boldsymbol{F}\dfrac{\partial W(\boldsymbol{C})}{\partial \boldsymbol{C}}\boldsymbol{F}^T \ , \ \boldsymbol{\sigma} = \dfrac{1}{J}\boldsymbol{\tau}$
4. referential elasticity tensor (AD): $\mathbb{C} = 4\dfrac{\partial^2 W(\boldsymbol{C})}{\partial \boldsymbol{C} \partial \boldsymbol{C}}$
5. push-forward of $\mathbb{C}$ to get spatial elasticity tensor: $\mathbb{c}_{ijkl} = F_{iI} F_{jJ} F_{kK} F_{lL} \mathbb{C}_{IJKL}$
6. spatial elasticity tensor related to Jaumann stress rate $\tilde{\mathbb{c}}^{\sigma}$:

$$\tilde{\mathbb{c}}_{ijkl} = \mathbb{c}_{ijkl} + \frac{1}{2}(\delta_{ik}\tau_{jl} + \tau_{ik}\delta_{jl} + \delta_{il}\tau_{jk} + \tau_{il}\delta_{jk}) \ , \ \tilde{\mathbb{c}}^{\sigma} = \frac{1}{J}\tilde{\mathbb{c}}$$

Box 6.1

Application of AceGen

The corresponding AceGen commands are given in Box 6.2 to show the implementation process. First, the right Cauchy-Green tensor must be computed. In the following the derivative of the strain energy function with respect to the right Cauchy-Green tensor $\boldsymbol{C}$ is computed and thus the auxiliary variable $\mathbb{C}$ introduced by AceGen is not allowed to change anymore. The signature of these variables must be "unique" which is accomplished by the command `SMSFreeze`. In this function, the tensor product to obtain $\boldsymbol{C}$ from the deformation gradient $\boldsymbol{F}$, represented by the symbolic variable $\mathbb{F}$, is performed as well. The optional parameter `"Symmetric"→True` indicates that the "freezed" symbolic tensor is symmetric which is important for a minimum amount of auxiliary variables. To compute the Kirchhoff stress the command for the automatic differentiation (AD) is `SMSD` which is used for the derivative of `W` with respect to $\mathbb{C}$. If necessary the AceGen specific operator ⊨ introduces new auxiliary variables which are optimized with respect to the code generation process. Then, it is possible to compute the Cauchy stress using the AceGen function `SMSDet` for the determinant of a tensor. The elasticity tensor follows directly from the second derivative with respect to the right Cauchy-Green tensor by using AD. The push-forward as well as the summation to get the spatial elasticity tensor (items 5 and 6 in Box 6.1) are performed by do loops in Mathematica syntax which is not depicted in Box 6.2 but is given in App. C.1 within the presentation of the complete code.

AceGen/Mathematica code

```
SMSFreeze[ℂ, Transpose[𝔽].𝔽, "Symmetric" → True];
dWdC ⊨ SMSD[W, ℂ, "Symmetric" → True];
τ ⊨ 2 * 𝔽 . dWdC . Transpose[𝔽];
J ⊨ SMSDet[𝔽];
σ ⊨ τ / J;
C ⊨ 4 * SMSD[dWdC, ℂ, "Symmetric" → True];
```

Box 6.2

Verification

The neo-Hookean strain energy function (6.7) is used to verify the computation of the stress and the automatic differentiation. For this purpose, the AceGen generated routine is used in a single element test and the simulation of a three point bending test. The stress, number of iterations and the convergence rate are compared with the solution based on the analytically calculated tangent. This "reference" solution is programmed with AceGen as well in order to get a comparable code, yet without performing any derivatives. The size of each routine in bytes is shown in Tab. 6.1.

Table 6.1.: By AceGen generated routines for a UMAT

analytical tangent	reference solution	2938 bytes
tangent by AceGen	AD (Box 6.1)	25813 bytes

Single element test In order to verify the AceGen computed tangent a uniaxial tensile test with one element is performed. In this test high strains ($\lambda_1 = 2.5$) are applied to the element which allows to check if the computation of the tangent is correct in a non-linear computation. The relative error η of the computed tangent with respect to the reference tangent ${}^{ref}\tilde{\mathbb{C}}^{\sigma}$ is evaluated for this purpose using the Frobenius norm

$$\eta = \left| \frac{\|\tilde{\mathbb{C}}^{\sigma}\| - \|{}^{ref}\tilde{\mathbb{C}}^{\sigma}\|}{\|{}^{ref}\tilde{\mathbb{C}}^{\sigma}\|} \right| \quad . \tag{6.12}$$

The material parameters in (6.7) are $C_{10} = 19.23$ MPa and $D_1 = 0.024\,\text{MPa}^{-1}$. A large load step is specified to force the algorithm to iterate more than two times to find a converged solution. The error is evaluated in the first iteration. The automatic differentiation method (AD) produces a very small error $\eta \approx 10^{-16}$ which is in the region of the machine precision of a 64bit architecture ($2.2 \cdot 10^{-16}$). The computation with the reference tangent and the AD tangent need six iterations to converge. The convergence behavior in terms of the norm of the largest residual force (largest unequilibrated nodal force) is identical (Tab. 6.2). The computation of the stress at the end of the simulation gives same values as well.

Table 6.2.: Norm of largest residual force

iteration #	analytical tangent	AD tangent
1	$4.447 \cdot 10^4$	$4.447 \cdot 10^4$
2	$1.961 \cdot 10^4$	$1.961 \cdot 10^4$
3	$7.741 \cdot 10^3$	$7.741 \cdot 10^3$
4	815	815
5	7.27	7.27
6	$6.554 \cdot 10^{-4}$	$6.554 \cdot 10^{-4}$

Simulation of three point bending test Additionally to the single element test a three point bending simulation of a rubber plate serves as a second computational example with a large number of degree of freedoms. Again, the number of iterations, convergence rate and the computation times are compared as well as the errors of the tangent with respect to the reference tangent. Additionally, it serves as an example to compare the computation time in case of a large number of degrees of freedom. In this simulation, contact occurs between the plate and a die (pushing the plate down) and between the plate and the support, respectively. Hence a highly non-linear mechanical boundary value problem must be solved. The rubber plate is modeled with the dimensions: length=100mm, depth=50mm, thickness=3mm. The die and the support are modeled as rigid bodies and the plate is discretized with about 21000 3d hexahedral linear reduced integration elements with a denser mesh in the region of possible contact (Fig. 6.1a). The max. displacement of the die is 10 mm (Fig. 6.1b).

(a) undeformed plate (b) deformed plate

Figure 6.1.: FE-model of three point bending simulation

In both simulations, the total number of iterations are the same and a quadratic convergence rate of the norm of the residual force is obtained near the solution (Tab. 6.3). The convergence behavior shows that the tangent is computed correctly. The computation time (Tab. 6.3) is slightly higher in case of the AD due to more operations performed per iteration. This correlates of course with the size of the routines (Tab. 6.1).

Table 6.3.: Convergence behavior in simulation of three point bending test. The norm of the largest residual force is given in the third increment.

	analytical tangent	AD tangent
total no. of iterations	22	22
normalized computation wall-clock time	1.0	1.0278
iteration #		
1	14.9	14.9
2	5.41	5.41
3	$1.705 \cdot 10^{-2}$	$1.705 \cdot 10^{-2}$
4	$8.687 \cdot 10^{-6}$	$8.687 \cdot 10^{-6}$

6.2.2. Elastic-plastic material model

In this section the implementation of an elastic-plastic material model with AceGen is presented. As in Sec. 6.2.1 the Cauchy stress and the consistent tangent moduli must be calculated. In case of plasticity this is much more complicated because a local Newton-scheme must be performed for the integration of the kinematics and the internal variables. The algorithmic consistent tangent then depends on the update algorithm (Hughes and Pister (1978), Simo and Taylor (1985)) and may be calculated by AceGen in a very smart way.

For simplicity and to show the main features of the implementation a standard isotropic elastic-plastic model with linear isotropic hardening and isothermal conditions is considered, e.g. Wriggers (2008), Neto et al. (2008), Lubliner (2008). An exponential map update scheme for the kinematics is used as in the models in Ch. 4. The constitutive equations and the numerical treatment is briefly summarized in the following.

As in Ch. 4, the deformation gradient is split into an elastic and a plastic part. The specific free energy Ψ is formulated in terms of the elastic right Cauchy-Green tensor and the accumulated plastic strain ε^p (isotropic hardening), and is assumed to be additively composed of an elastic and plastic part

$$\Psi(\hat{\boldsymbol{C}}^e, \varepsilon^p) = \Psi^e(\hat{\boldsymbol{C}}^e) + \Psi^p(\varepsilon^p) \tag{6.13}$$

with $\hat{\boldsymbol{C}}^e = \boldsymbol{F}^{eT}\boldsymbol{F}^e$. Performing the time derivatives of Ψ

$$\dot{\Psi}^e = \frac{\partial \Psi^e}{\partial \hat{\boldsymbol{C}}^e} : \dot{\hat{\boldsymbol{C}}}^e = 2\boldsymbol{F}^e \frac{\partial \Psi^e}{\partial \hat{\boldsymbol{C}}^e}\boldsymbol{F}^{eT} : \boldsymbol{d}^e \tag{6.14}$$

$$\dot{\Psi}^p = \frac{\partial \Psi^p}{\partial \varepsilon^p}\dot{\varepsilon}^p = q\dot{\varepsilon}^p \tag{6.15}$$

with the scalar quantity $q = \dfrac{\partial \Psi^p}{\partial \varepsilon^p}$ and inserting it in (2.94) yields the inequality

$$(\boldsymbol{\tau} - 2\rho_0 \boldsymbol{F}^e \frac{\partial \Psi^e}{\partial \hat{\boldsymbol{C}}^e}\boldsymbol{F}^{eT}) : \boldsymbol{d} + 2\rho_0 \boldsymbol{F}^e \frac{\partial \Psi^e}{\partial \hat{\boldsymbol{C}}^e}\boldsymbol{F}^{eT} : \boldsymbol{d}^p - \rho_0 q \dot{\varepsilon}^p \geq 0 \tag{6.16}$$

from which the Kirchhoff stress follows as (cf. Ch. 4)

$$\boldsymbol{\tau} = 2\rho_0 \boldsymbol{F}^e \frac{\partial \Psi^e}{\partial \hat{\boldsymbol{C}}^e}\boldsymbol{F}^{eT} = \boldsymbol{F}^e \hat{\boldsymbol{S}} \boldsymbol{F}^{eT} \tag{6.17}$$

with the second Piola-Kirchhoff stress tensor in the intermediate configuration

$$\hat{\boldsymbol{S}} = 2\rho_0 \frac{\partial \Psi^e}{\partial \hat{\boldsymbol{C}}^e} \quad . \tag{6.18}$$

The reduced form of (6.16) yields the internal dissipation due to plastic deformation

$$\mathcal{D}_{int} = \boldsymbol{\tau} : \boldsymbol{d}^p - \rho_0 q \dot{\varepsilon}^p \geq 0 \quad . \tag{6.19}$$

which provides a restriction on the material parameters to ensure the thermodynamic consistency of the material model. The principle of maximum dissipation gives the flow rule for the plastic part of the rate of deformation tensor and the evolution equation of the internal variable ε^p (Lubliner, 2008)

$$\boldsymbol{d}^p = \lambda \boldsymbol{n} \quad , \quad \boldsymbol{n} = \frac{\partial f}{\partial \boldsymbol{\tau}} \quad , \quad \dot{\varepsilon}^p = \lambda h \quad , \quad h = -\frac{\partial f}{\partial q} \tag{6.20}$$

with the plastic multiplier λ, the flow function $f(\boldsymbol{\tau}, q) \leq 0$ and the loading/unloading (Karush-Kuhn-Tucker) conditions

$$\lambda \geq 0 \quad , \quad f \leq 0 \quad , \quad \lambda f = 0 \quad . \tag{6.21}$$

In the following, the isotropic von Mises yield criterion is used. Since this criterion was developed for ductile metals only the deviatoric part of the Kirchhoff stress tensor

$$\boldsymbol{\tau}' = \boldsymbol{\tau} - \frac{1}{3}\mathrm{tr}[\boldsymbol{\tau}]\mathbf{1} \tag{6.22}$$

is accounted for in the yield function because hydrostatic stresses lead to negligible plastic deformations. The flow function reads

$$f(\boldsymbol{\tau}, q) = \tau_m - \sqrt{\frac{2}{3}}(\tau_0 + q) \tag{6.23}$$

with $\tau_m = \mathrm{tr}[\boldsymbol{\tau}'\boldsymbol{\tau}']^{1/2}$ and the initial yield stress τ_0. The plastic potential is chosen to be quadratic

$$W^p = \frac{1}{2}k(\varepsilon^p)^2 \tag{6.24}$$

from which with (6.15) the scalar $q = k\varepsilon^p$ is obtained and where k is a material parameter representing the hardening modulus. The elastic potential is chosen to be the same neo-Hookean strain energy function as in Sec. 6.2.1 but is formulated as a function of $\hat{\boldsymbol{C}}^e$.

A further derivation of the second Piola-Kirchhoff stress tensor in the intermediate configuration with respect to the elastic right Cauchy-Green tensor leads to the intermediate elasticity tensor

$$\hat{\mathbb{C}}^e = 2\frac{\partial \hat{\boldsymbol{S}}}{\partial \hat{\boldsymbol{C}}^e} \quad . \tag{6.25}$$

Numerical treatment

In order to compute the stress, the standard elastic-predictor plastic-corrector algorithm is used (Simo and Ortiz (1985), Wriggers (2008), Neto et al. (2008), Shabana (2008), Hashiguchi and Yamakawa (2012)). In the predictor (or trial step), the elastic deformation is set by

$$\boldsymbol{F}^{e,tr} = \boldsymbol{F}_{t+\Delta t}\boldsymbol{F}_t^{p-1} \tag{6.26}$$

to compute the trial stress with which the flow function is tested whether the deformation state is elastic or plastic. If the material is still in the elastic regime, the flow function is less than zero, the new stress is equal to the trial stress and the consistent tangent is simply the elasticity tensor (see (6.4)). But, if the flow function is greater than zero, which is actually not admissible, plastic deformation occurres and the corrector step must be employed with the so-called radial return mapping scheme by projecting the stress back on the yield surface so that $f(\boldsymbol{\tau}, q) = 0$ holds. In that case the internal variables must be integrated. For this purpose, the differential equation

$$\dot{\boldsymbol{F}}^p = \hat{\boldsymbol{l}}^p \boldsymbol{F}^p \tag{6.27}$$

which reduces with the assumption of $\hat{\boldsymbol{w}}^p = \boldsymbol{0}$ to

$$\dot{\boldsymbol{F}}^p = \hat{\boldsymbol{d}}^p \boldsymbol{F}^p \tag{6.28}$$

is approximated by an exponential function, the exponential map, in the time interval $[t, t + \Delta t]$ (Simo, 1992)

$$\boldsymbol{F}^p_{t+\Delta t} = \exp[\Delta t\, \hat{\boldsymbol{d}}^p_{t+\Delta t}] \boldsymbol{F}^p_t \quad . \tag{6.29}$$

The above equation may be reformulated in terms of the elastic deformation gradient and by using (6.20) it yields

$$\boldsymbol{F}^e_{t+\Delta t} = \exp[-\Delta\lambda\, \boldsymbol{n}_{t+\Delta t}] \boldsymbol{F}^{e,tr} \tag{6.30}$$

with $\Delta\lambda = \Delta t \dot{\lambda}$. The evolution equation for the accumulated plastic strain is integrated by a backward Euler-scheme

$$\varepsilon^p_{t+\Delta t} = \varepsilon^p_t + \Delta t\, \dot{\varepsilon}^p = \varepsilon^p_t + \Delta\lambda\, h_{t+\Delta t} \tag{6.31}$$

and the Karush-Kuhn-Tucker conditions (6.21) in incremental form read

$$\Delta\lambda \geq 0 \quad , \quad f_{t+\Delta t} \leq 0 \quad , \quad \Delta\lambda f_{t+\Delta t} = 0 \quad . \tag{6.32}$$

This strongly non-linear system of equations is merged in the residual vector $\boldsymbol{R}$

$$\boldsymbol{R} = [\boldsymbol{R}_1\ R_2\ R_3]^T = \boldsymbol{0} \tag{6.33}$$

with

$$\boldsymbol{R}_1 = \boldsymbol{F}^e_{t+\Delta t} - \exp[-\Delta\lambda\, \boldsymbol{n}_{t+\Delta t}] \boldsymbol{F}^{e,tr} \tag{6.34}$$

$$R_2 = \varepsilon^p_{t+\Delta t} - \varepsilon^p_t - \Delta\lambda\, h_{t+\Delta t} \tag{6.35}$$

$$R_3 = f(\boldsymbol{\tau}_{t+\Delta t}, q_{t+\Delta t}) \quad . \tag{6.36}$$

It can be solved by a Newton-scheme for the vector of unknown variables $\boldsymbol{X} = [\boldsymbol{F}^p \; \Delta\lambda \; \varepsilon^p]^T$. For this purpose a first-order Taylor series approximation is considered at a certain known state $\boldsymbol{R}_i$

$$\boldsymbol{R}_{i+1} = \boldsymbol{R}_i + \frac{\partial \boldsymbol{R}_i}{\partial \boldsymbol{X}_i} \Delta \boldsymbol{X} = \boldsymbol{0} \tag{6.37}$$

which is solved for the increment of the unknowns

$$\Delta \boldsymbol{X} = -\boldsymbol{J}^{-1} \, * \, \boldsymbol{R}_i \tag{6.38}$$

where $\boldsymbol{J} = \left(\frac{\partial \boldsymbol{R}}{\partial \boldsymbol{X}}\right)$ is the Jacobian. The symbol $*$ denotes the appropriate contraction. The unknown vector is then updated by

$$\boldsymbol{X}_{i+1} = \boldsymbol{X}_i + \Delta \boldsymbol{X} \tag{6.39}$$

which is done in an iteration loop until the scheme has converged.

The tangent consistent with the integration scheme is obtained by the derivative (Hughes and Pister (1978), Simo and Taylor (1985), Simo and Hughes (1998))

$$\mathbb{C}^{alg} = 2\frac{\partial \boldsymbol{S}}{\partial \boldsymbol{C}} \tag{6.40}$$

and may be pushed forward to get the tangent in terms of quantities in the spatial configuration (e.g. Holopainen and Wallin (2013))

$$\mathbb{c}^{alg}_{ijkl} = \frac{\partial \tau_{ij}}{\partial F_{km}} F_{lm} - (\tau_{il}\delta_{jk} + \tau_{jl}\delta_{ik}) \quad . \tag{6.41}$$

With (6.4) one gets the consistent tangent related to the Cauchy stress needed in Abaqus/-Standard

$$\tilde{\mathbb{c}}^{alg,\sigma}_{ijkl} = J^{-1}\left[\frac{\partial \tau_{ij}}{\partial F_{km}} F_{lm} + \frac{1}{2}(\delta_{il}\tau_{jk} + \tau_{ik}\delta_{jl} - \delta_{ik}\tau_{jl} - \tau_{il}\delta_{jk})\right] \quad . \tag{6.42}$$

The implicit derivatives that arise in the tangent may be obtained by extracting them from the Jacobian of the already performed Newton-scheme (Johansson et al. (2005), Ekh and Runesson (2001)). Supposing that

$$\boldsymbol{R}(\boldsymbol{X}(\boldsymbol{F}), \boldsymbol{F}) = \boldsymbol{0} \; \Rightarrow \; \mathrm{d}\boldsymbol{R} = \boldsymbol{0} \tag{6.43}$$

holds for a converged state, we have

$$\mathrm{d}\boldsymbol{R} = \left(\frac{\partial \boldsymbol{R}}{\partial \boldsymbol{X}} * \frac{\partial \boldsymbol{X}}{\partial \boldsymbol{F}} + \frac{\partial \boldsymbol{R}}{\partial \boldsymbol{F}}\right) : \mathrm{d}\boldsymbol{F} = \boldsymbol{0} \tag{6.44}$$

from which the required derivatives may be obtained by using the inverse of the Jacobian

$$\frac{\partial \boldsymbol{X}}{\partial \boldsymbol{F}} = -\boldsymbol{J}^{-1} * \frac{\partial \boldsymbol{R}}{\partial \boldsymbol{F}} \quad . \tag{6.45}$$

In Box 6.3 the complete algorithm is demonstrated; again all points are marked where the automatic differentiation (AD) is used. The algorithm is formulated in a general way that an implementation using AceGen is efficiently applicable. The elastic ($W^e = \rho_0 \Psi^e$) and plastic ($W^p = \rho_0 \Psi^p$) potential as well as the flow function f are not specified so that the algorithm is flexible in terms of exchanging material models.

Algorithm to compute the stress and the consistent tangent moduli

input in UMAT: $\boldsymbol{F}_{t+\Delta t}$, $\boldsymbol{F}^p_t$, ε^p_t

→ computation with AceGen :

1. **trial step:**

   1.1 trial elastic deformation gradient $\boldsymbol{F}^{e,tr} = \boldsymbol{F}_{t+\Delta t} \boldsymbol{F}^{p-1}_t$

   1.2 trial stress (AD): $\boldsymbol{\tau}^{tr} = 2\boldsymbol{F}^{e,tr} \dfrac{\partial W^e}{\partial \hat{\boldsymbol{C}}^{e,tr}} \boldsymbol{F}^{eT,tr}$

   1.3 check if $f(\boldsymbol{\tau}^{tr}, q) < 0$ with $q = \dfrac{\partial W^p}{\partial \varepsilon^p_t}$ (AD)

   if true: convergence, compute new stress and elasticity tensor

   $$\boldsymbol{\tau}_{t+\Delta t} = \boldsymbol{\tau}^{tr}, \quad \boldsymbol{F}^p_{t+\Delta t} = \boldsymbol{F}^p_t, \quad \varepsilon^p_{t+\Delta t} = \varepsilon^p_t, \quad \hat{\mathbb{C}}^e = 2\frac{\partial \hat{\boldsymbol{S}}^{tr}}{\partial \hat{\boldsymbol{C}}^{e,tr}} \quad \text{(AD)}$$

   $$\tilde{\mathbb{C}}^{alg,\sigma}_{ijkl} = J^{-1}[F^e_{iI} F^e_{jJ} F^e_{kK} F^e_{lL} \hat{\mathbb{C}}^e_{IJKL} + \frac{1}{2}(\delta_{ik}\tau_{jl} + \tau_{ik}\delta_{jl} + \delta_{il}\tau_{jk} + \tau_{il}\delta_{jk})]$$

   END

   else: go to 2

2. **radial return mapping scheme:**

   2.1 Newton iteration loop: $i = 0, 1, ...,$ convergence

   I. initialization, set: $\boldsymbol{F}^p_0 = \boldsymbol{F}^p_t$, $\Delta\lambda_0 = 0$, $\varepsilon^p_0 = \varepsilon^p_t$

II. $\boldsymbol{F}^e_i = \boldsymbol{F}_{t+\Delta t}\boldsymbol{F}^{p-1}_i$, $\boldsymbol{\tau}_i = 2\boldsymbol{F}^e_i \dfrac{\partial W^e_i}{\partial \hat{\boldsymbol{C}}^e_i}\boldsymbol{F}^{eT}_i$, $\boldsymbol{n}_i = \dfrac{\partial f_i}{\partial \boldsymbol{\tau}_i}$, $q_i = \dfrac{\partial W^p_i}{\partial \varepsilon^p_i}$, $h_i = -\dfrac{\partial f_i}{\partial q_i}$

(AD)

III. compute residual vector: $\boldsymbol{R}_i = [\boldsymbol{R}_{1i}\ R_{2i}\ R_{3i}]^T$ with

$$\boldsymbol{R}_{1i} = \boldsymbol{F}^e_i - \exp[-\Delta\lambda_i\, \boldsymbol{n}_i]\boldsymbol{F}^{e,tr},\ R_{2i} = \varepsilon^p_i - \varepsilon^p_t - \Delta\lambda_i h_i,\ R_{3i} = f_i$$

IV. solve (AD): $\Delta \boldsymbol{X} = -\left(\dfrac{\partial \boldsymbol{R}_i}{\partial \boldsymbol{X}_i}\right)^{-1} * \boldsymbol{R}_i \rightarrow \boldsymbol{X}_{i+1} = \boldsymbol{X}_i + \Delta \boldsymbol{X} = [\boldsymbol{F}^p_{i+1}\ \Delta\lambda_{i+1}\ \varepsilon^p_{i+1}]^T$

V. check convergence criterion:

$$\|\Delta \boldsymbol{X}\| \begin{cases} \geq \text{Tol} \rightarrow \text{set } i = i+1 \text{ and go to II.} \\ < \text{Tol} \rightarrow \text{ convergence : solve (AD) } \dfrac{\partial \boldsymbol{X}_{i+1}}{\partial \boldsymbol{F}_{t+\Delta t}} = -\left(\dfrac{\partial \boldsymbol{R}_{i+1}}{\partial \boldsymbol{X}_{i+1}}\right)^{-1} * \dfrac{\partial \boldsymbol{R}_{i+1}}{\partial \boldsymbol{F}_{t+\Delta t}} \\ \quad \text{set : } \boldsymbol{F}^p_{t+\Delta t} = \boldsymbol{F}^p_{i+1},\ \varepsilon^p_{t+\Delta t} = \varepsilon^p_{i+1}, \text{ END loop and go to 2.2} \end{cases}$$

2.2 compute new stress: $\boldsymbol{F}^e_{t+\Delta t} = \boldsymbol{F}_{t+\Delta t}\boldsymbol{F}^{p-1}_{t+\Delta t}$, $\quad \boldsymbol{\tau}_{t+\Delta t} = 2\boldsymbol{F}^e_{t+\Delta t}\dfrac{\partial W^e_{t+\Delta t}}{\partial \hat{\boldsymbol{C}}^e_{t+\Delta t}}\boldsymbol{F}^{eT}_{t+\Delta t}$ (AD)

2.3 compute consistent tangent (AD):

$$\tilde{\mathbb{C}}^{alg,\sigma}_{ijkl} = J^{-1}\left[\frac{\partial \tau_{ij}}{\partial F_{km}}F_{lm} + \frac{1}{2}(\delta_{il}\tau_{jk} + \tau_{ik}\delta_{jl} - \delta_{ik}\tau_{jl} - \tau_{il}\delta_{jk})\right]$$

Box 6.3

In the next section the application of AceGen is shown. A special treatment is necessary to obtain the derivatives for the consistent tangent.

Application of AceGen

The realization of the algorithm of Box 6.3 with AceGen is shown in Box 6.4 and Box 6.5 and is described with the crucial functions in the following.

First, in Box 6.4 the trial step is performed by computing the trial stress and evaluating the flow function. The functions of the potentials `We` and `Wp` and the flow function `f` are kept empty to indicate that specific and appropriate functions may be inserted. That means, `f` must be a function of $\boldsymbol{\tau}$, `We` a function of `ℂe` and `Wp` of `ϵpt0`. The history variables $\boldsymbol{F}^p_t$ and ε^p_t must be treated carefully by saving them at the beginning in the field `Ht0` and passing them to the field of the history variables Ht1 at the end of the time step or to the iteration field X1 depending on the result of the evaluation of the flow function after the trial step. The former case occurs if the flow function is less than zero. This is checked by the if-else statement SMSIf[SMSLogical[ ]] and the function producing a logical expression.

AceGen and Mathematica code of trial step

```
𝔽etr ⊨ 𝔽t1.SMSInverse[𝔽pt0];
SMSFreeze[ℂe, Transpose[𝔽etr].𝔽etr,
  "Ignore" → (# === 0 &), "Symmetric" → True];
We ⊨ ...;
τtrial ⊨ 2 *𝔽etr.SMSD[We, ℂe, "IgnoreNumbers" → True,
     "Symmetric" → True].Transpose[𝔽etr];
Wp ⊨ ...;
q ⊨ SMSD[Wp, ϵpt0];
f ⊨ ...;

Ht0 ⊨ ({𝔽pt0⟦1, 1⟧, 𝔽pt0⟦1, 2⟧, 𝔽pt0⟦1, 3⟧, 𝔽pt0⟦2, 1⟧, 𝔽pt0⟦2, 2⟧,
    𝔽pt0⟦2, 3⟧, 𝔽pt0⟦3, 1⟧, 𝔽pt0⟦3, 2⟧, 𝔽pt0⟦3, 3⟧, Δλt0, ϵpt0});

SMSIf[SMSLogical[f < 10^-8]];
Ht1 ⊣ Ht0;
```

Box 6.4

If the expression is not true the return-mapping algorithm starts by applying a Newton-scheme which is depicted in Box 6.5. First, the history field is restored in the field Xi which includes the variables to solve for and changes during the iteration. This iteration loop is initialized by SMSDo[ ] in which Xi must be passed in. The components of Xi are saved to the temporary tensor field 𝔽pi and the scalars Δλi and ϵpi. After the computation of the stress and the flow function the residual vectors are assembled and (6.30) is coded by using the AceGen function SMSMatrixExp[ ] for an exponential function of tensors. The derivative of the residuum with respect to the history variables is obtained by AD and the unknown vector ΔX is solved by performing a LU-factorization (SMSLUFactor[ ]) with a subsequent Gauss elimination procedure (SMSLUSolve[ ]). The solution increment is added to the history variable field with the operator for multi-valued auxiliary variables

⊣ and the convergence is checked using a stopping criterion. If the increment of growth is sufficiently small the iteration is stopped. In that case, the derivative of the history variables with respect to the deformation gradient is computed for the use in the setup of the consistent tangent. This is realized by computing first the derivative of the residual vector with respect to the deformation gradient and additionally by holding the vector of the converged vector of unknowns fixed by applying the option `"Constant"→Xi` in the `SMSD` command. This is important because the whole iteration scheme would be considered for the computation of the derivative instead of using the converged state of the variables only. A further Gauss-elimination procedure gives the desired derivative. The converged variables are then stored in the field `Ht1` and the dependency of the previously computed derivative (6.45) is stated for the computation of the consistent tangent moduli. At the end of the time step the plastic deformation gradient is assembled from which the elastic deformation gradient and the stress is computed in the following code. The consistent tangent moduli is computed by AD in case of elasticity with item 1.3 and in case of plasticity with item 2.3 of Box 6.3 which is then simply accomplished by the `SMSD` function because all implicit derivatives and dependencies are known. The complete code is given in App. C.2.

Verification

Analogous to the hyperelastic material model in Sec. 6.2.1, the implementation of the elastic-plastic model is first verified by a single element test. Afterwards, a uniaxial tensile test of a cylindrical bar is simulated. It should be noted that no analytical closed solution for the consistent tangent is derived nor is implemented in this work. It will be verified by making use of a numerically approximated tangent. Two different approximation methods will be used for this purpose (cf. App. B).

Single element test The aim of the single element test is to compare the convergence behavior of the tangent obtained with the AD method with numerically approximated tangents. The finite difference method (FD) and the complex step derivative approximation method (CSDA) are used for the numerical approximation of the tangent (see. App. B). The additional material parameters are chosen as $\tau_0 = 300$ MPa and $k = 20$ MPa. The convergence parameters are adjusted so that the computation needs three increments with 11 iterations in total to converge to the solution. By comparing the convergence behavior it is found that the norm of the largest residual force is the same for the AD computed tangent and the CSDA tangent (Tab. 6.4). See App. B for a detailed discussion of the two numerical approximation methods.

AceGen and Mathematica code of return mapping algorithm

```
SMSElse[];
Xi ⊣ Ht0;
SMSDo[ii, 1, 30, 1, Xi];
 𝔽pi ⊨ ({{Xi⟦1⟧, Xi⟦2⟧, Xi⟦3⟧}, {Xi⟦4⟧, Xi⟦5⟧, Xi⟦6⟧},
     {Xi⟦7⟧, Xi⟦8⟧, Xi⟦9⟧}});
 Δλi ⊨ Xi⟦10⟧;
 SMSFreeze[ϵpi, Xi⟦11⟧];
 𝔽ei ⊨ 𝔽t1.SMSInverse[𝔽pi];
 SMSFreeze[ℭei, Transpose[𝔽ei].𝔽ei, "Symmetric" → True];
 Wei ⊨ ...;
 SMSFreeze[τi, 2 * 𝔽ei.SMSD[Wei, ℭei, "IgnoreNumbers" → True,
       "Symmetric" → True].Transpose[𝔽ei], "KeepStructure" → True];
 Wpi ⊨ ...;
 SMSFreeze[qi, SMSD[Wpi, ϵpi]];
 fi ⊨ ...;
 ni ⊨ Simplify[SMSD[fi, τi, "IgnoreNumbers" → True,
     "Symmetric" → True]];
 hi ⊨ -SMSD[fi, qi];
 R1i ⊨ Simplify[𝔽ei - SMSMatrixExp[-Δλi * ni, "Order" → 10,
       "Module" → False].𝔽etr];
 R2i ⊨ ϵpi - ϵpt0 - Δλi * hi;
 R3i ⊨ fi;
 Ri ⊨ {R1i⟦1, 1⟧, R1i⟦1, 2⟧, R1i⟦1, 3⟧, R1i⟦2, 1⟧, R1i⟦2, 2⟧,
    R1i⟦2, 3⟧, R1i⟦3, 1⟧, R1i⟦3, 2⟧, R1i⟦3, 3⟧, R2i, R3i};
 RX ⊨ SMSD[Ri, Xi];
 LU ⊣ SMSLUFactor[RX];
 ΔX ⊨ SMSLUSolve[LU, -Ri];
 Xi ⊣ Xi + ΔX;
 ΔXN ⊨ SMSSqrt[ΔX.ΔX / Length[ΔX]];
 SMSIf[ΔXN < 10 ^ -8];
  dRd𝔽 ⊨ SMSD[Ri, Flatten[𝔽t1], "Constant" → Xi];
  dXd𝔽 ⊣ SMSLUSolve[LU, -dRd𝔽];
  SMSBreak[];
 SMSEndIf[];
SMSEndDo[Xi, dXd𝔽];
Ht1 ⊣ SMSFreeze[Xi, "Dependency" → {Flatten[𝔽t1], dXd𝔽}];
SMSEndIf[Ht1];
𝔽pt1 ⊨ ({{Ht1⟦1⟧, Ht1⟦2⟧, Ht1⟦3⟧}, {Ht1⟦4⟧, Ht1⟦5⟧, Ht1⟦6⟧},
     {Ht1⟦7⟧, Ht1⟦8⟧, Ht1⟦9⟧}});
ϵpt1 ⊨ Ht1⟦11⟧;
```

Box 6.5

Table 6.4.: Norm of largest residual force in the first increment.

iteration #	AD	FD	CSDA
1	$1.271 \cdot 10^3$	$1.271 \cdot 10^3$	$1.271 \cdot 10^3$
2	10.4	10.4	10.4
3	$6.787 \cdot 10^{-4}$	$6.798 \cdot 10^{-4}$	$6.787 \cdot 10^{-4}$

Simulation of a cylindrical bar In order to test the computed tangent in a more sophisticated computation a uniaxial tensile test of a cylindrical bar is simulated (Simo, 1992). The bar is geometrically imperfect to induce necking. For this purpose, the radius $r_0 = 5$ mm decreases linear to a radius in the center of the bar of $r_m = 0.98 \cdot r_0$ mm. Because of symmetry only one-eighth of the bar is modeled with 128764 second-order tetrahedron elements (C3D10, 555504 dof). At one end the bar is fixed while at the other end a displacement boundary condition is applied with a total displacement of 6 mm at an initial length of 50 mm of the complete bar. Furthermore, symmetry boundary conditions are considered. In Fig. 6.2 the initial full undeformed model of the bar and the deformed bar with the neck is shown.

(a) undeformed bar (b) deformed bar

Figure 6.2.: FE-model of the bar

All three methods to compute the tangent (AD, FD, CSDA) result in the same convergence behavior (Tab. 6.5). They all need the same number of iterations and the convergence rate of the residual force is the same as well.

6.3. Conclusions

In this chapter, the package AceGen is illustrated as a tool to implement constitutive equations and their numerical treatment in terms of computing the stress and the algorithmic consistent tangent moduli. This is necessary for the user material model interface

Table 6.5.: Convergence behavior of the simulation. The norm of the largest residual force is given in the last increment.

	AD tangent (AceGen)	FD tangent	CSDA tangent
total no. of iterations	227	227	227
iteration #			
1	0.430	0.430	0.430
2	0.128	0.128	0.128
3	$3.606 \cdot 10^{-2}$	$3.606 \cdot 10^{-2}$	$3.606 \cdot 10^{-2}$
4	$9.177 \cdot 10^{-3}$	$9.177 \cdot 10^{-3}$	$9.177 \cdot 10^{-3}$
5	$1.570 \cdot 10^{-3}$	$1.569 \cdot 10^{-3}$	$1.569 \cdot 10^{-3}$

(UMAT) of the implicit solver of Abaqus/Standard. In both considered material models, the hyperelastic and the elastic-plastic model, the derivatives are computed using the automatic differentiation feature of AceGen. This is verified in case of the hyperelastic model by the analytical solution for the tangent and in case of the elastic-plastic model by two different approximation methods of the consistent tangent. The error of the automatic differentiation method is in the range of the machine precision compared to the analytical solution of the hyperelastic model and the convergence behavior is the same. It could be shown that the convergence behavior of simulations using the elastic-plastic model and the automatic differentiated tangent is identical to the tangents obtained by the numerical approximation methods. Thus, the tangent generated by AceGen could be verified and can be used for a UMAT. However, this required a new approach of implementation using AceGen within the Mathematica environment which brings a rather long familiarization, respectively, development time. If this barrier is overcome the implementation is much more easy and not error-prone like in a standard programming language, for example FORTRAN. If the basic structure is implemented the constitutive equations are exchangeable and a completely new material behavior including the consistent tangent is programmed in minutes. This makes AceGen highly valuable in the process of material modeling because different approaches may be tested without spending a lot of time for programming and calculation of derivatives.

7. Conclusions & Outlook

7.1. Conclusions

The purpose of the present work was to develop a constitutive model for amorphous thermoplastic polymers. The main focus laid on the consideration of a wide range of temperatures - from room temperature till above the glass transition temperature - and to account for a pre-oriented molecular network as it may arise from manufacturing processes.

For this purpose, an existing model for glassy polymers by Boyce et al. (1988) was enhanced to be also suitable at elevated temperatures. The intermolecular and molecular network resistances against inelastic deformation of this basic model were modified to capture the strong temperature dependence. The modification of the intermolecular resistance - particularly Young's modulus and the shear yield strength - led to a very good agreement with own experimental results in the small strain regime. For the molecular network resistance two different models were incorporated into the basic model: the "entanglement dissociation model" and the "molecular relaxation by reptation" model for which the formulation of an enhanced kinematics must be performed. Both models were able to reproduce the temperature dependence of the hardening behavior at large strains of an amorphous thermoplastic polymer (PMMA) with introducing only a limited number of additional material parameters. The stress response of the models was compared to stress-strain curves obtained from own uniaxial tensile tests conducted from room temperature till above the glass transition temperature at different strain rates. In the experiments the local strain field (utilized also to calculate the true stress) was successfully measured using digital image correlation.

In analyzing a fictitious non-monotonic thermomechanical loading process the entanglement dissociation model failed and an unrealistic stress response was obtained. In contrast, the molecular relaxation (reptation) model gave a realistic stress response. Thus, this model is much more suitable for simulations of thermoforming, for example, where complicated thermomechanical loading is applied to the material.

The molecular relaxation (reptation) model was further validated by simulating micro-thermoforming processes. Experiments with different process parameters were simulated and the computed deformations were compared to the real formed parts. In the simulations, the cooling and mold opening stages were also considered which is a rare approach in literature but is important when it comes to further structural simulations with the final component. In that case the stress and strain distribution in the component are thus available. Despite several simplifications the computational model captured the influence of different process parameters quite well and the simulated forming displacements were in a very good agreement with experimentally obtained forming shapes. This showed that the developed constitutive model can be successfully used for thermoforming process simulations.

The incorporation of a pre-orientation of the molecular network into the constitutive models could be successfully accomplished as well. An initial network stretch tensor was introduced for this purpose which contains information about the pre-deformation history of the molecular network, e.g. from processing. This affects the large strain response of the material and could be qualitatively reproduced by the model. The deformation behavior of two injection molded components at heating above the glass transition temperature was investigated and served as computational examples. By mapping an assumed initial molecular orientation on the finite element model, the extended constitutive model was successfully used to simulate the re-heating process. The resulting deformation in the simulations of both parts are in good agreement with that observed in the real experiment. In a further computational example, a thermoforming process of a blister-like part was simulated and different amounts of initially applied pre-stretch in the polymer film were investigated. It could be shown that an initial stretch in the film has a strong influence on the thermoforming behavior. With increasing pre-stretch the resistance against forming increased and the forming depth decreased. Thus, the pre-orientation of the molecular network should be regarded in thermoforming simulations.

The developed material models were implemented into the FE-program Abaqus. For this purpose the automatic differentiation, code generation and optimization tool AceGen was utilized. Since Acegen is a plug-in for the computer algebra system Mathematica, symbolic programming of the constitutive equations could be used for the implementation which simplifies the code development. However, to use this tool efficiently and to take full advantage of the capabilities a different programming approach had to be developed, e.g. in computing the algorithmic tangent. The generated code was verified on two simpler constitutive models - a hyperelastic and an elastic-plastic model - by comparing the algorithmic tangents with analytical and numerically approximated tangents. In summary, AceGen could be used successfully and efficiently to implement user material models in Abaqus. By using this tool, it is not necessary anymore to form complicated derivatives to obtain the algorithmic tangent. Rather, one can implement changes in the material model very quickly and obtain the tangent automatically.

7.2. Outlook

Although the molecular relaxation (reptation) model proved to be in good agreement with the experimental results, the model response can be further improved. For instance, consideration of adiabatic heating at higher strain rates could give a better response at lower temperatures. This can be easily included into the model as shown in Arruda et al. (1995), for example. The intrinsic softening of the material at lower ambient temperatures (deliberately ignored in the present work) could be incorporated as well to give a better fit to the experiments. However, at higher temperatures the softening effect vanishes and thus is not important for thermoforming analyses. A further improvement of the model would be to include the temperature and strain rate dependent viscoelasticity which is important for the cyclic (unloading) behavior and which was not investigated at all in this work.

In the simulations of micro-thermoforming processes the boundary conditions could be refined to be more realistic, for example, to clamp the film accounting for friction and not to simply fix it at the boundary. In addition, a thermal analysis prior to the mechanical analysis could be performed so that then the computed temperature distribution could serve as a temperature boundary condition. This could further improve the quality of the computational model and the results. In a further step, a coupled temperature-displacement analysis could bring even more improvements. Moreover, easy to perform experiments that mimic a thermoforming process could be used to validate the constitutive model without introducing several unknowns. This could be performed e.g. in uniaxial tensile tests with non-constant displacement and temperature conditions. Similar experiments on extruded polymer rods with different amounts of applied stretch were conducted by Kahar et al. (1978). These data could also be used to validate and to improve a constitutive model.

To obtain realistic values for the pre-stretch in a polymer a methodology should be worked out, for example, with measurements of the birefringence or an estimation obtained from melt flow simulations.

Bibliography

Adams, A. M., Buckley, C. P., and Jones, D. P. (2000). Biaxial hot drawing of poly(ethylene terephthalate): measurements and modelling of strain-stiffening. *Polymer*, 41(2):771–786.

Agah-Tehrani, A., Lee, E. H., Mallett, R. L., and Onat, E. T. (1987). The theory of elastic-plastic deformation at finite strain with induced anisotropy modeled as combined isotropic-kinematic hardening. *Journal of the Mechanics and Physics of Solids*, 35(5):519–539.

Ames, N. M., Srivastava, V., Chester, S. A., and Anand, L. (2009). A thermo-mechanically coupled theory for large deformations of amorphous polymers. part ii: Applications. *International Journal of Plasticity*, 25(8):1495 – 1539.

Anand, L. and Gurtin, M. E. (2003). A theory of amorphous solids undergoing large deformations, with application to polymeric glasses. *International Journal of Solids and Structures*, 40(6):1465 – 1487.

Argon, A. S. (1973). A theory for the low-temperature plastic deformation of glassy polymers. *Philosophical Magazine*, 28(4):839–865.

Arruda, E. M. and Boyce, M. C. (1993). A three-dimensional constitutive model for the large stretch behavior of rubber elastic materials. *Journal of the Mechanics and Physics of Solids*, 41(2):389–412.

Arruda, E. M., Boyce, M. C., and Jayachandran, R. (1995). Effects of strain rate, temperature and thermomechanical coupling on the finite strain deformation of glassy polymers. *Mechanics of Materials*, 19(2–3):193–212.

Arruda, E. M., Boyce, M. C., and Quintus-Bosz, H. (1993). Effects of initial anisotropy on the finite strain deformation behavior of glassy polymers. *International Journal of Plasticity*, 9(7):783–811.

aus der Wiesche, S. (2004). Industrial thermoforming simulation of automotive fuel tanks. *Applied Thermal Engineering*, 24(16):2391 – 2409.

Becker, F. (2009). *Entwicklung einer Beschreibungsmethodik für das mechanische Verhalten unverstärkter Thermoplaste bei hohen Deformationsgeschwindigkeiten*. Dissertation, Martin-Luther-Universität Halle-Wittenberg.

Belytschko, T., Liu, W. K., and Moran, B. (2000). *Nonlinear Finite Elements for Continua and Structures*. Wiley.

Bergström, J. S. and Boyce, M. C. (1998). Constitutive modeling of the large strain time-dependent behavior of elastomers. *Journal of the Mechanics and Physics of Solids*, 46(5):931–954.

Boyce, M. C., Parks, D. M., and Argon, A. S. (1988). Large inelastic deformation of glassy polymers. part i: rate dependent constitutive model. *Mechanics of Materials*, 7(1):15–33.

Boyce, M. C., Parks, D. M., and Argon, A. S. (1989a). Plastic flow in oriented glassy polymers. *International Journal of Plasticity*, 5(6):593–615.

Boyce, M. C., Socrate, S., and Llana, P. G. (2000). Constitutive model for the finite deformation stress–strain behavior of poly(ethylene terephthalate) above the glass transition. *Polymer*, 41(6):2183–2201.

Boyce, M. C., Weber, G. G., and Parks, D. M. (1989b). On the kinematics of finite strain plasticity. *Journal of the Mechanics and Physics of Solids*, 37(5):647–665.

Carlone, P. and Palazzo, G. (2006). Finite element analysis of the thermoforming manufacturing process using the hyperelastic mooney-rivlin model. In Gavrilova, M., Gervasi, O., Kumar, V., Tan, C., Taniar, D., Laganá, A., Mun, Y., and Choo, H., editors, *Computational Science and Its Applications - ICCSA 2006*, volume 3980 of *Lecture Notes in Computer Science*, pages 794–803. Springer Berlin Heidelberg.

Chen, K. and Schweizer, K. S. (2009). Suppressed segmental relaxation as the origin of strain hardening in polymer glasses. *Physical review letters*, 102(3):038301. PMID: 19257400.

Cohen, D. A. (1991). A padé approximant to the inverse langevin function. *Rheologica Acta*, 30(3):270–273.

Coleman, B. D. and Gurtin, M. E. (1967). Thermodynamics with internal state variables. *The Journal of Chemical Physics*, 47(2):597–613.

Dantec Dynamics A/S (2012). *Istra4D manual*.

Dassault Systèmes (2012). *Abaqus Documentation 6.12*. http://www.simulia.com.

De Focatiis, D. S. A. and Buckley, C. P. (2011). Prediction of frozen-in birefringence in oriented glassy polymers using a molecularly aware constitutive model allowing for finite molecular extensibility. *Macromolecules*, 44(8):3085–3095.

de Gennes, P. G. (1971). Reptation of a polymer chain in the presence of fixed obstacles. *The Journal of Chemical Physics*, 55(2):572–579.

de Gennes, P.-G. (1979). *Scaling Concepts in Polymer Physics*. Cornell University Press.

de Gennes, P.-G. (1983). Entangled polymers. *Physics Today*, 36(6):33–39.

Doi, M. and Edwards, S. F. (1978). Dynamics of concentrated polymer systems. part 1.—Brownian motion in the equilibrium state. *Journal of the Chemical Society, Faraday Transactions 2: Molecular and Chemical Physics*, 74(0):1789–1801.

Dooling, P., Buckley, C., Rostami, S., and Zahlan, N. (2002a). Hot-drawing of poly(methyl methacrylate) and simulation using a glass–rubber constitutive model. *Polymer*, 43(8):2451 – 2465.

Dooling, P. J., Buckley, C. P., Rostami, S., and Zahlan, N. (2002b). Hot-drawing of poly(methyl methacrylate) and simulation using a glass–rubber constitutive model. *Polymer*, 43(8):2451–2465.

Dupaix, R. B. and Boyce, M. C. (2007). Constitutive modeling of the finite strain behavior of amorphous polymers in and above the glass transition. *Mechanics of Materials*, 39(1):39–52.

Ekh, M. and Runesson, K. (2001). Modeling and numerical issues in hyperelasto-plasticity with anisotropy. *International Journal of Solids and Structures*, 38(52):9461–9478.

Engelmann, S. (2012). *Advanced Thermoforming: Methods, Machines and Materials, Applications and Automation*. Wiley Series on Polymer Engineering and Technology. Wiley.

Govaert, L. E. and Tervoort, T. A. (2004). Strain hardening of polycarbonate in the glassy state: Influence of temperature and molecular weight. *Journal of Polymer Science Part B: Polymer Physics*, 42(11):2041–2049.

G'Sell, C., Hiver, J. M., Dahoun, A., and Souahi, A. (1992). Video-controlled tensile testing of polymers and metals beyond the necking point. *Journal of Materials Science*, 27(18):5031–5039.

Gurtin, M. E., Fried, E., and Anand, L. (2010). *The Mechanics and Thermodynamics of Continua*. Cambridge.

Hasan, O. A. and Boyce, M. C. (1995). A constitutive model for the nonlinear viscoelastic viscoplastic behavior of glassy polymers. *Polymer Engineering & Science*, 35(4):331–344.

Hasan, O. A., Boyce, M. C., Li, X. S., and Berko, S. (1993). An investigation of the yield and postyield behavior and corresponding structure of poly(methyl methacrylate). *Journal of Polymer Science Part B: Polymer Physics*, 31(2):185–197.

Hashiguchi, K. and Yamakawa, Y. (2012). *Introduction to Finite Strain Theory for Continuum Elasto-Plasticity*. John Wiley & Sons.

Haward, R. N. and Thackray, G. (1968). The use of a mathematical model to describe isothermal stress-strain curves in glassy thermoplastics. *Royal Society of London Proceedings Series A*, 302:453–472.

Haward, R. N. and Young, R. (1997). *The Physics of Glassy Polymers*. Springer.

Heilig, M. (2012). private communication.

Heilig, M., Schneider, M., Ide, T., Dinglreiter, H., and Worgull, M. (2010). Technology of Micro Thermoforming and Micro Blow Moulding for the Integration of Selective Surfacemodifications in three-dimensional Microstructures. *Proceedings of the Polymer Processing Society 26th Annual Meeting, PPS-26, July 4-8, 2010 Banff (Canada)*.

Holopainen, S. and Wallin, M. (2013). Modeling of the long-term behavior of glassy polymers.

Holzapfel, G. A. (2007). *Nonlinear Solid Mechanics*. Wiley.

Hoy, R. S. and Robbins, M. O. (2006). Strain hardening of polymer glasses: Effect of entanglement density, temperature, and rate. *Journal of Polymer Science Part B: Polymer Physics*, 44(24):3487–3500.

Hoy, R. S. and Robbins, M. O. (2008). Strain hardening of polymer glasses: Entanglements, energetics, and plasticity. *Physical Review E*, 77(3).

Hughes, T. J. and Pister, K. S. (1978). Consistent linearization in mechanics of solids and structures. *Computers & Structures*, 8(3–4):391–397.

Instron (2012). *Wavematrix manual*.

Intel (2013). *Intel Fortran Compiler*.
`http://software.intel.com/en-us/fortran-compilers`.

Johansson, G., Ekh, M., and Runesson, K. (2005). Computational modeling of inelastic large ratcheting strains. *International Journal of Plasticity*, 21(5):955–980.

Kahar, N., Duckett, R. A., and Ward, I. M. (1978). Stress optical studies of oriented poly(methyl methacrylate). *Polymer*, 19(2):136–144.

Kamlah, M. (1994). *Zur Modellierung des Verfestigungsverhaltens von Materialien mit statischer Hysterese im Rahmen der phänomenologischen Thermomechanik*. Dissertation, Institut für Mechanik, Universität Gesamthochschule Kassel.

Karamanou, M., Warby, M., and Whiteman, J. (2006). Computational modelling of thermoforming processes in the case of finite viscoelastic materials. *Computer Methods in Applied Mechanics and Engineering*, 195(37–40):5220 – 5238. John H. Argyris Memorial Issue. Part I.

Kim, G., Lee, K., and Kang, S. (2009). Prediction of the film thickness distribution and pattern change during film insert thermoforming. *Polymer Engineering & Science*, 49(11):2195–2203.

Kinloch, A. J. and Young, R. J. (1983). *Fracture Behaviour of Polymers*. Applied Science Publishers.

Klompen, E. T. J., Engels, T. A. P., Govaert, L. E., and Meijer, H. E. H. (2005). Modeling of the postyield response of glassy polymers: Influence of thermomechanical history. *Macromolecules*, 38(16):6997–7008.

Korelc, J. (1996). *Symbolic approach in computational mechanics and its application to the enhanced strain method*. Dissertation, Instiut für Mechanik, TH Darmstadt.

Korelc, J. (2002). Multi-language and multi-environment generation of nonlinear finite element codes. *Eng. Comput.*, 18(4):312–327.

Korelc, J. (2009). *AceGen*.
`http://www.fgg.uni-lj.si/Symech/User/AceGenManual.pdf`.

Korelc, J. (2012). *AceGen & AceFEM*. `http://www.fgg.uni-lj.si/Symech/`.

Korelc, J. and Wriggers, P. (1999). Computer algebra and automatic differentiation in derivation of finite element code. *ZAMM - Zeitschrift für Angewandte Mathematik und Mechanik*, 79:811–812.

Kotlik, M. (2011). *Experimentelle Charakterisierung eines thermoplastischen Kunststoffes bei großen Deformationen*. Diplomarbeit, Institut für Mechanik, Karlsruhe Institut für Technologie (KIT).

Kouba, K., Bartos, O., and Vlachopoulos, J. (1992). Computer simulation of thermoforming in complex shapes. *Polymer Engineering & Science*, 32(10):699–704.

Kuhn, W. and Grün, F. (1942). Beziehungen zwischen elastischen konstanten und dehnungsdoppelbrechung hochelastischer stoffe. *Kolloid-Zeitschrift*, 101(3):248–271.

Lee, E. H. (1969). Elastic-plastic deformation at finite strains. *Journal of Applied Mechanics*, 36(1):1–6.

LSTC (2013a). *LS-OPT 5.0*. `http://www.lstc.com/`.

LSTC (2013b). *LS-OPT 5.0 User's Manual*.

Lubliner, J. (2008). *Plasticity Theory*. Dover Publications.

Makradi, A., Ahzi, S., Gregory, R., and Edie, D. (2005). A two-phase self-consistent model for the deformation and phase transformation behavior of polymers above the glass transition temperature: application to {PET}. *International Journal of Plasticity*, 21(4):741 – 758.

Makradi, A., Belouettar, S., Ahzi, S., and Puissant, S. (2007). Thermoforming process of amorphous polymeric sheets: Modeling and finite element simulations. *Journal of Applied Polymer Science*, 106(3):1718–1724.

Martins, J. R. R. A., Sturdza, P., and Alonso, J. J. (2003). The complex-step derivative approximation. *ACM Trans. Math. Softw.*, 29(3):245–262.

Mathworks (2013). *Matlab R2013a*. `http://www.mathworks.de/products/matlab/`.

Mattern, S. (2012). *Hocheffiziente Formulierung und Implementierung finiter Elemente für transiente Analysen mit expliziter Zeitintegration*. Dissertation, Institut für Mechanik, Karlsruhe Institut für Technologie (KIT).

McCool, R. and Martin, P. J. (2011). Thermoforming process simulation for the manufacture of deep-draw plastic food packaging. *Proceedings of the Institution of Mechanical Engineers, Part E: Journal of Process Mechanical Engineering*, 225(4):269–279.

Miehe, C. (1996). Numerical computation of algorithmic (consistent) tangent moduli in large-strain computational inelasticity. *Computer Methods in Applied Mechanics and Engineering*, 134:223 – 240.

Miehe, C., Diez, J. M., Göktepe, S., and Schänzel, L.-M. (2011). Coupled thermoviscoplasticity of glassy polymers in the logarithmic strain space based on the free volume theory. *International Journal of Solids and Structures*, 48(13):1799 – 1817.

Miehe, C., Göktepe, S., and Lulei, F. (2004). A micro-macro approach to rubber-like materials—part i: the non-affine micro-sphere model of rubber elasticity. *Journal of the Mechanics and Physics of Solids*, 52(11):2617 – 2660.

Nam, G. J., Lee, J. W., and Ahn, K. H. (2000). Three-dimensional simulation of thermoforming process and its comparison with experiments. *Polymer Engineering & Science*, 40(10):2232–2240.

Neto, E. d. S., Peric, D., and Owens, D. (2008). *Computational Methods for Plasticity*. Wiley.

OpenAD (2013). `http://www.mcs.anl.gov/OpenAD/`.

O'Connor, C., Martin, P., Sweeney, J., Menary, G., Caton-Rose, P., and Spencer, P. (2013). Simulation of the plug-assisted thermoforming of polypropylene using a large strain thermally coupled constitutive model. *Journal of Materials Processing Technology*, 213(9):1588 – 1600.

Palm, G. O. and Dupaix, R. (2006). Large strain material characterization and modeling of poly (methyl methacrylate) (pmma) near the glass transition. *J. Eng. Mater. Technol*, 128(4):559–563.

Pan, B., Qian, K., Xie, H., and Asundi, A. (2009). Two-dimensional digital image correlation for in-plane displacement and strain measurement: a review. *Measurement Science and Technology*, 20(6):062001.

Parisch, H. (2003). *Festkörper-Kontinuumsmechanik*. Teubner.

Perez-Foguet, A., Rodriguez-Ferran, A., and Huerta, A. (2000). Numerical differentiation for local and global tangent operators in computational plasticity. *Computer Methods in Applied Mechanics and Engineering*, 189(1):277–296.

Polanco-Loria, M., Clausen, A. H., Berstad, T., and Hopperstad, O. S. (2010). Constitutive model for thermoplastics with structural applications. *International Journal of Impact Engineering*, 37(12):1207 – 1219.

Raha, S. and Bowden, P. B. (1972). Birefringence of plastically deformed poly(methyl methacrylate). *Polymer*, 13(4):174–183.

Reese, S. (2003). A micromechanically motivated material model for the thermo-viscoelastic material behaviour of rubber-like polymers. *International Journal of Plasticity*, 19(7):909 – 940.

Richeton, J., Ahzi, S., Vecchio, K., Jiang, F., and Makradi, A. (2007a). Modeling and validation of the large deformation inelastic response of amorphous polymers over a wide range of temperatures and strain rates. *International Journal of Solids and Structures*, 44(24):7938 – 7954.

Richeton, J., Ahzi, S., Vecchio, K. S., Jiang, F. C., and Makradi, A. (2007b). Modeling and validation of the large deformation inelastic response of amorphous polymers over a wide range of temperatures and strain rates. *International Journal of Solids and Structures*, 44(24):7938–7954.

Rösler, J., Harders, H., and Bäker, M. (2003). *Mechanisches Verhalten der Werkstoffe: mit 28 Tabellen und 31 Aufgaben*. Teubner, Stuttgart; Leipzig; Wiesbaden.

Sagar, G. and Stein, E. (2008). Convergence behavior of 3-d finite elements for neo-hookean material models using abaqus-umat. *Int. J. for Compputer-Aided Engng. and Software*, 25(3):220–232.

Senden, D. J. A., Krop, S., van Dommelen, J. a. W., and Govaert, L. E. (2012). Rate- and temperature-dependent strain hardening of polycarbonate. *Journal of Polymer Science Part B: Polymer Physics*, 50(24):1680–1693.

Senden, D. J. A., van Dommelen, J. a. W., and Govaert, L. E. (2010). Strain hardening and its relation to bauschinger effects in oriented polymers. *Journal of Polymer Science Part B: Polymer Physics*, 48(13):1483–1494.

Shabana, A. A. (2008). *Computational continuum mechanics*. Cambridge University Press, Cambridge; New York.

Simo, J. (1992). Algorithms for static and dynamic multiplicative plasticity that preserve the classical return mapping schemes of the infinitesimal theory. *Computer Methods in Applied Mechanics and Engineering*, 99(1):61–112.

Simo, J. and Ortiz, M. (1985). A unified approach to finite deformation elastoplastic analysis based on the use of hyperelastic constitutive equations. *Computer Methods in Applied Mechanics and Engineering*, 49(2):221–245.

Simo, J. and Taylor, R. (1985). Consistent tangent operators for rate-independent elasto-plasticity. *Computer Methods in Applied Mechanics and Engineering*, 48(1):101–118.

Simo, J. C. and Hughes, T. J. R. (1998). *Computational Inelasticity*. Springer.

Srivastava, V., Chester, S. A., Ames, N. M., and Anand, L. (2010). A thermo-mechanically-coupled large-deformation theory for amorphous polymers in a temperature range which spans their glass transition. *International Journal of Plasticity*, 26(8):1138–1182.

Strobl, G. R. (2007). *The Physics of Polymers: Concepts for Understanding Their Structures and Behavior*. Springer.

Struik, L. C. E. (1990). *Internal stresses, dimensional instabilities, and molecular orientations in plastics*. Wiley.

Sun, W., Chaikof, E. L., and Levenston, M. E. (2008). Numerical approximation of tangent moduli for finite element implementations of nonlinear hyperelastic material models. *Journal of Biomechanical Engineering*, 130(6):061003.

Sutton, M. A., Orteu, J. J., and Schreier, H. (2009). *Image Correlation for Shape, Motion and Deformation Measurements: Basic Concepts,Theory and Applications*. Springer.

Tanaka, M., Fujikawa, M., Balzani, D., and Schröder, J. (2014). Robust numerical calculation of tangent moduli at finite strains based on complex-step derivative approximation and its application to localization analysis. *Computer Methods in Applied Mechanics and Engineering*, 269:454–470.

Tervoort, T., Smit, R., Brekelmans, W., and Govaert, L. (1997). A constitutive equation for the elasto-viscoplastic deformation of glassy polymers. *Mechanics of Time-Dependent Materials*, 1(3):269–291.

Tomita, Y. and Tanaka, S. (1995). Prediction of deformation behavior of glassy polymers based on molecular chain network model. *International Journal of Solids and Structures*, 32(23):3423 – 3434.

Treloar, L. R. G. (1974). *Introduction to polymer science*. Wykeham Publications ; Springer-Verlag, London; New York.

van Breemen, L., Klompen, E., Govaert, L., and Meijer, H. (2011). Extending the {EGP} constitutive model for polymer glasses to multiple relaxation times. *Journal of the Mechanics and Physics of Solids*, 59(10):2191 – 2207.

Wang, M. C. and Guth, E. (1952). Statistical theory of networks of nongaussian flexible chains. *The Journal of Chemical Physics*, 20(7):1144–1157.

Warby, M., Whiteman, J., Jiang, W.-G., Warwick, P., and Wright, T. (2003). Finite element simulation of thermoforming processes for polymer sheets. *Mathematics and Computers in Simulation*, 61(3–6):209 – 218. {MODELLING} 2001 - Second {IMACS} Conference on Mathematical Modelling and Computational Methods in Mechanics, Physics, Biomechanics and Geodynamics.

Ward, I. M. and Sweeney, J. (2005). *An Introduction to the Mechanical Properties of Solid Polymers*. John Wiley & Sons.

Wendlandt, M., Tervoort, T. A., and Suter, U. W. (2005). Non-linear, rate-dependent strain-hardening behavior of polymer glasses. *Polymer*, 46(25):11786–11797.

Williams, M. L., Landel, R. F., and Ferry, J. D. (1955). The temperature dependence of relaxation mechanisms in amorphous polymers and other glass-forming liquids. *Journal of the American Chemical Society*, 77(14):3701–3707.

Wolfram (2013). *Mathematica 9*. `http://www.wolfram.com/mathematica/`.

Wriggers, P. (2008). *Nonlinear Finite Element Methods*. Springer.

Wu, P. and van der Giessen, E. (1994). Analysis of shear band propagation in amorphous glassy polymers. *International Journal of Solids and Structures*, 31(11):1493 – 1517.

Wu, P. D. and van der Giessen, E. (1993). On improved network models for rubber elasticity and their applications to orientation hardening in glassy polymers. *Journal of the Mechanics and Physics of Solids*, 41(3):427–456.

Appendix

A. Notes to the constitutive models of Ch. 4

A.1. Material parameters

Table A.1.: Material parameter values for PMMA in the basic model for PMMA (Sec. 4.1.2)

E [MPa]	2500	$N(T = 293.15\text{K}) = N_{RT}$ [−]	2
ν [−]	0.4	$C^R(T = 293.15\text{K}) = C^R_{RT}$ [MPa]	5
$\dot{\gamma}^p_0$ [s^{-1}]	$1.707 \cdot 10^{25}$		
A [MPa K^{-1}]	131		
s_0 [MPa]	180		
α [−]	0.2		

Table A.2.: Material parameter values for PMMA in the entanglement dissociation model according to (Raha and Bowden, 1972) (Sec. 4.2.1)

T_g [K]	378.15	A_T [K^{-1}]	0.00323
ΔT_g [K]	15	s_0 [MPa]	180
E_{g1} [MPa]	9096	α [−]	0.2
E_{g2} [MPa K^{-1}]	22.5	N [−]	2
E_{r1} [MPa]	100	C^R [MPa]	5
E_{r2} [MPa K^{-1}]	4.15	B [mm^{-3}]	$2.9547 \cdot 10^{18}$
ν [−]	0.4	D [mm^{-3}]	$1.20087 \cdot 10^{22}$
$\dot{\gamma}^p_0$[s^{-1}]	$1 \cdot 10^{26}$	E_a [kcal mol^{-1}]	6.543
A_0 [MPa^{-1}]	1.213		

Table A.3.: Material parameter values for PMMA in the molecular relaxation by reptation model (Sec. 4.2.2)

T_g [K]	378.15	α [−]	0.2
ΔT_g [K]	15	N [−]	2
E_{g1} [MPa]	9096	C^R [MPa]	5
E_{g2} [MPa K^{-1}]	22.5	$\dot{\gamma}^p_{N0}$ [s^{-1}]	$3.3 \cdot 10^{23}$
E_{r1} [MPa]	100	A_{NT} [K]	20541
E_{r2} [MPa]	4.15	κ [−]	6.756
ν [−]	0.4	λ^L_{N0} [−]	$3.9 \cdot 10^{-11}$
$\dot{\gamma}^p_0$[s^{-1}]	$1 \cdot 10^{26}$	λ^L_{NT} [K^{-1}]	0.06306
A_0 [MPa^{-1}]	1.213	ξ [−]	55
A_T [K^{-1}]	0.00323	$\tilde{\tau}^b_0$ [MPa]	1
s_0 [MPa]	180		

A.2. Numerical treatment of material models of Ch. 4

In this section the numerical treatment with respect to the implementation using AceGen of the models of the Ch. 4 is shown. For the implementation using AceGen, especially the treatment of the Newton iteration scheme and the computation of the algorithmic tangent, it is referred to App. C.2 where the complete code of the implementation of a elastic-plastic material model is presented. The implementation of the models of Ch. 4 are very similar to the model presented in App. C.2.

A.2.1. Basic model of Sec. 4.1.2

In contrast to the standard von Mises plasticity model the glassy polymer model is not formulated with a flow function. Thus, the algorithm to compute the new stress is straight forward and no elastic-predictor plastic-corrector algorithm is necessary. The kinematics and the plastic strain are updated with a Newton-scheme in which the plastic deformation gradient is integrated using the exponential mapping approximation

$$\boldsymbol{F}^p_{t+\Delta t} = \exp[\Delta t\, \hat{\boldsymbol{l}}^p_{t+\Delta t}]\boldsymbol{F}^p_t \tag{A.1}$$

and the plastic strain with an Euler backward method

$$\gamma^p_{t+\Delta t} = \gamma^p_t + \Delta t \dot{\gamma}^p \quad . \tag{A.2}$$

Setting the plastic spin to zero ($\hat{\boldsymbol{w}}^p = \boldsymbol{0}$) (A.1) yields

$$\boldsymbol{F}^p_{t+\Delta t} = \exp[\Delta t\, \hat{\boldsymbol{d}}^p_{t+\Delta t}]\boldsymbol{F}^p_t \quad . \tag{A.3}$$

The resulting non-linear system of equations is combined in the residual vector $\boldsymbol{R}$

$$\boldsymbol{R} = [\boldsymbol{R}_1\ R_2]^T = \boldsymbol{0} \tag{A.4}$$

with

$$\boldsymbol{R}_1 = \boldsymbol{F}^p_{t+\Delta t} - \exp[\Delta t\, \hat{\boldsymbol{d}}^p_{t+\Delta t}]\boldsymbol{F}^p_t \tag{A.5}$$

$$R_2 = \gamma^p_{t+\Delta t} - \gamma^p_t - \Delta t \dot{\gamma}^p \quad . \tag{A.6}$$

This is solved by the Newton-scheme as shown in Sec. 6.2.2 for the unknown variables vector $\boldsymbol{X} = [\boldsymbol{F}^p\ \gamma^p]^T$. The algorithm is shown in Box A.1.

Algorithm to compute the stress and the consistent tangent moduli for the basic model

input in UMAT: $\boldsymbol{F}_{t+\Delta t}, \boldsymbol{F}^p_t, \gamma^p_t$

$\rightarrow$ computation with AceGen :

1 Newton iteration loop: $i = 0, 1, ...,$ convergence

I. initialization, set: $\boldsymbol{F}^p_0 = \boldsymbol{F}^p_t,\ \gamma^p_0 = \gamma^p_t$

II. compute (AD): $\boldsymbol{F}^e_i = \boldsymbol{F}_{t+\Delta t}\boldsymbol{F}^{p-1}_i,\ \boldsymbol{\Sigma}_i = 2\hat{\boldsymbol{C}}^e_i \dfrac{\partial W^e_i}{\partial \hat{\boldsymbol{C}}^e_i},\ \hat{\boldsymbol{\tau}}^b_i = \dfrac{C^R \beta_i}{3\lambda_{CLi}} \hat{\boldsymbol{b}}^p_i,\ \hat{\boldsymbol{d}}^p_i = \dot{\gamma}^p_i \dfrac{\boldsymbol{\Sigma}^{*\prime}_i}{\|\boldsymbol{\Sigma}^{*\prime}_i\|}$

III. compute residual vector: $\boldsymbol{R}_i = [\boldsymbol{R}_{1i}\ R_{2i}]^T$ with

$\boldsymbol{R}_{1i} = \boldsymbol{F}^p_i - \exp[\Delta t\ \hat{\boldsymbol{d}}^p_i]\boldsymbol{F}^p_t,\ R_{2i} = \gamma^p_i - \gamma^p_t - \Delta t \dot{\gamma}^p_i$

IV. solve (AD): $\Delta \boldsymbol{X} = -\left(\dfrac{\partial \boldsymbol{R}_i}{\partial \boldsymbol{X}_i}\right)^{-1} * \boldsymbol{R}_i \ \rightarrow\ \boldsymbol{X}_{i+1} = \boldsymbol{X}_i + \Delta \boldsymbol{X} = [\boldsymbol{F}^p_{i+1}\ \gamma^p_{i+1}]^T$

V. check convergence criterion:

$$
\|\Delta \boldsymbol{X}\| \begin{cases} \geq \text{Tol} \rightarrow \text{set } i = i+1 \text{ and go to II.} \\ < \text{Tol} \rightarrow \ \text{convergence : solve (AD) } \dfrac{\partial \boldsymbol{X}_{i+1}}{\partial \boldsymbol{F}_{t+\Delta t}} = -\left(\dfrac{\partial \boldsymbol{R}_{i+1}}{\partial \boldsymbol{X}_{i+1}}\right)^{-1} * \dfrac{\partial \boldsymbol{R}_{i+1}}{\partial \boldsymbol{F}_{t+\Delta t}} \\ \quad \text{set : } \boldsymbol{F}^p_{t+\Delta t} = \boldsymbol{F}^p_{i+1},\ \gamma^p_{t+\Delta t} = \gamma^p_{i+1}, \text{ END loop and go to 2} \end{cases}
$$

2 compute new stress (AD): $\boldsymbol{F}^e_{t+\Delta t} = \boldsymbol{F}_{t+\Delta t}\boldsymbol{F}^{p-1}_{t+\Delta t},\quad \boldsymbol{\tau}_{t+\Delta t} = 2\boldsymbol{F}^e_{t+\Delta t} \dfrac{\partial W^e_{t+\Delta t}}{\partial \hat{\boldsymbol{C}}^e_{t+\Delta t}} \boldsymbol{F}^{eT}_{t+\Delta t}$

3 compute consistent tangent (AD):

$$
\tilde{\mathbb{C}}^{alg,\sigma}_{ijkl} = J^{-1}\left[\frac{\partial \tau_{ij}}{\partial F_{km}} F_{lm} + \frac{1}{2}(\delta_{il}\tau_{jk} + \tau_{ik}\delta_{jl} - \delta_{ik}\tau_{jl} - \tau_{il}\delta_{jk})\right]
$$

Box A.1

A.2.2. Molecular relaxation by reptation model of Sec. 4.2.2

Due to the extended kinematics of the basic model of Sec. 4.1.2 the numerical costs increase in case of the molecular relaxation model. For the Newton-scheme nine additional equations are to solve. The residual vector then yields

$$\boldsymbol{R} = [\boldsymbol{R}_1 \; \boldsymbol{R}_1 \; R_2]^T = \boldsymbol{0} \tag{A.7}$$

with

$$\boldsymbol{R}_1 = \boldsymbol{F}^p_{t+\Delta t} - \exp[\Delta t \; \hat{\boldsymbol{d}}^p_{t+\Delta t}]\boldsymbol{F}^p_t \tag{A.8}$$

$$\boldsymbol{R}_2 = \boldsymbol{F}^p_{Nt+\Delta t} - \exp[\Delta t \; \tilde{\boldsymbol{d}}^p_{Nt+\Delta t}]\boldsymbol{F}^p_{Nt} \tag{A.9}$$

$$R_3 = \gamma^p_{t+\Delta t} - \gamma^p_t - \Delta t \dot{\gamma}^p \quad . \tag{A.10}$$

The overall algorithm is given in Box A.2.

Algorithm to compute the stress and the consistent tangent moduli for the reptation model

input in UMAT: $\boldsymbol{F}_{t+\Delta t}$, $\boldsymbol{F}^p_t$, $\boldsymbol{F}^p_{Nt}$,γ^p_t

→ computation with AceGen :

1 Newton iteration loop: $i = 0, 1, ...,$ convergence

I. initialization, set: $\boldsymbol{F}^p_0 = \boldsymbol{F}^p_t$, $\boldsymbol{F}^p_{N0} = \boldsymbol{F}^p_{Nt}$ $\gamma^p_0 = \gamma^p_t$

II. compute (AD): $\boldsymbol{F}^e_i = \boldsymbol{F}_{t+\Delta t}\boldsymbol{F}^{p-1}_i$, $\boldsymbol{F}^e_{Ni} = \boldsymbol{F}^p_i\boldsymbol{F}^{p-1}_{Ni}$, $\boldsymbol{\Sigma}_i = 2\hat{\boldsymbol{C}}^e_i\dfrac{\partial W^e_i}{\partial \hat{\boldsymbol{C}}^e_i}$, $\hat{\boldsymbol{\tau}}^b_i = \dfrac{C^R\beta_i}{3\lambda_{CLi}}\hat{\boldsymbol{b}}^e_{Ni}$

$$\hat{\boldsymbol{d}}^p_i = \dot{\gamma}^p_i\frac{\boldsymbol{\Sigma}^{*\prime}_i}{\|\boldsymbol{\Sigma}^{*\prime}_i\|},\ \tilde{\boldsymbol{d}}^p_{Ni} = \dot{\gamma}^p_{Ni}\frac{\tilde{\boldsymbol{\tau}}^{b\prime}_i}{\|\tilde{\boldsymbol{\tau}}^{b\prime}_i\|}$$

III. compute residual vector: $\boldsymbol{R}_i = [\boldsymbol{R}_{1i}\ \boldsymbol{R}_{2i}\ R_{3i}]^T$ with

$$\boldsymbol{R}_{1i} = \boldsymbol{F}^p_i - \exp[\Delta t\ \hat{\boldsymbol{d}}^p_i]\boldsymbol{F}^p_t,\ \boldsymbol{R}_{2i} = \boldsymbol{F}^p_{Ni} - \exp[\Delta t\ \tilde{\boldsymbol{d}}^p_{Ni}]\boldsymbol{F}^p_{Nt},\ R_{3i} = \gamma^p_i - \gamma^p_t - \Delta t\dot{\gamma}^p_i$$

IV. solve (AD): $\Delta\boldsymbol{X} = -\left(\dfrac{\partial \boldsymbol{R}_i}{\partial \boldsymbol{X}_i}\right)^{-1} * \boldsymbol{R}_i \ \rightarrow\ \boldsymbol{X}_{i+1} = \boldsymbol{X}_i + \Delta\boldsymbol{X} = [\boldsymbol{F}^p_{i+1}\ \boldsymbol{F}^p_{Ni+1}\ \gamma^p_{i+1}]^T$

V. check convergence criterion:

$$\|\Delta\boldsymbol{X}\|\begin{cases}\geq \text{Tol} \rightarrow \text{set } i = i+1 \text{ and go to II.}\\ < \text{Tol} \rightarrow\ \text{convergence : solve (AD) } \frac{\partial \boldsymbol{X}_{i+1}}{\partial \boldsymbol{F}_{t+\Delta t}} = -\left(\frac{\delta \boldsymbol{R}_{i+1}}{\partial \boldsymbol{X}_{i+1}}\right)^{-1} * \frac{\partial \boldsymbol{R}_{i+1}}{\partial \boldsymbol{F}_{t+\Delta t}}\\ \quad \text{set : } \boldsymbol{F}^p_{t+\Delta t} = \boldsymbol{F}^p_{i+1},\ \boldsymbol{F}^p_{Nt+\Delta t} = \boldsymbol{F}^p_{Ni+1},\ \gamma^p_{t+\Delta t} = \gamma^p_{i+1},\ \text{END loop and go to 2}\end{cases}$$

2 compute new stress (AD): $\boldsymbol{F}^e_{t+\Delta t} = \boldsymbol{F}_{t+\Delta t}\boldsymbol{F}^{p-1}_{t+\Delta t}$, $\quad \boldsymbol{\tau}_{t+\Delta t} = 2\boldsymbol{F}^e_{t+\Delta t}\dfrac{\partial W^e_{t+\Delta t}}{\partial \hat{\boldsymbol{C}}^e_{t+\Delta t}}\boldsymbol{F}^{eT}_{t+\Delta t}$

3 compute consistent tangent (AD):

$$\tilde{\mathbb{C}}^{alg,\sigma}_{ijkl} = J^{-1}\left[\frac{\partial \tau_{ij}}{\partial F_{km}}F_{lm} + \frac{1}{2}(\delta_{il}\tau_{jk} + \tau_{ik}\delta_{jl} - \delta_{ik}\tau_{jl} - \tau_{il}\delta_{jk})\right]$$

Box A.2

B. Numerical approximation methods of the tangent moduli

In this chapter the numerical approximation of the elasticity tensor and the algorithmic tangent is discussed. Again, AceGen is used for the programming of these tangents. The numerical approximation of the consistent tangent was first presented in Miehe (1996). An application for hyperelastic material models can be found for example in Sun et al. (2008) and Sagar and Stein (2008). The approximation in their works base on the forward difference approximation (FD). A furter approximation method is the complex step derivative approximation (CSDA) and is applied for plasticity regarding small deformations by Perez-Foguet et al. (2000) and hyperelasticity by Tanaka et al. (2014). Both approximation methods are briefly described in the following sections.

B.1. Numerical approximation of elasticity tensor

B.1.1. Forward difference approximation

The approximation of a derivative of a function $f(x)$ with the forward difference method bases on a small perturbation value ε. At a certain evaluation point the function value is perturbed and the difference to the unperturbed function is calculated. By dividing this difference by the perturbation value ε, the approximation of the derivative is as follows

$$\frac{\mathrm{d}f(x)}{\mathrm{d}x} \approx \frac{f(x+\varepsilon)-f(x)}{\varepsilon} \quad . \tag{B.1}$$

Applying this method to approximate the derivative necessary for the tangent moduli, the deformation gradient $\boldsymbol{F}$ must be perturbed in six directions (deformation states). So, the perturbed deformation gradient $\boldsymbol{F}^{\varepsilon}_{(ij)}$ reads

$$\boldsymbol{F}^{\varepsilon}_{(ij)} = \boldsymbol{F} + \Delta\boldsymbol{F}^{\varepsilon}_{(ij)} \tag{B.2}$$

with the perturbation

$$\Delta F^{\varepsilon}_{(ij)} = \frac{\varepsilon}{2}\left(e_i \otimes e_j F + e_j \otimes e_i F\right) \tag{B.3}$$

and the six choices for $e_{i,j=1,3}$ with $(i,j) = \{(1,1),(2,2),(3,3),(1,2),(1,3),(2,3)\}$ (Miehe, 1996). The increment of the Kirchhoff stress is then defined as the difference of the perturbed Kirchhoff stress $\tau^{\varepsilon}_{(ij)} = \tau(F^{\varepsilon}_{(ij)})$ and the unperturbed one $\tau = \tau(F)$ determined by the unperturbed deformation (cf. Box 6.1)

$$\Delta\tau^{\varepsilon}_{(ij)} = \tau^{\varepsilon}_{(ij)} - \tau \tag{B.4}$$

where the perturbed Kirchhoff stress must be computed six times for the six different perturbed deformation states. To get the approximated spatial tangent moduli $\tilde{\mathbb{c}}^{\varepsilon}_{(ij)}$ the linearization of the Kirchhoff stress in terms of the Jaumann-Zaremba stress rate is expressed in terms of the perturbed deformation gradient

$$\Delta\tau^{\varepsilon}_{(ij)} = \tilde{\mathbb{c}}^{\varepsilon}_{(ij)} : \Delta d^{\varepsilon}_{(ij)} + \Delta w^{\varepsilon}_{(ij)}\tau - \tau\Delta w^{\varepsilon}_{(ij)} \quad . \tag{B.5}$$

From the above equation it is obvious that the choice of only six perturbed deformation states is sufficient because all tensors are symmetric or antisymmetric, respectively. Furthermore, $\Delta w^{\varepsilon}_{(ij)}$ and $\Delta d^{\varepsilon}_{(ij)}$ depend on the incremental perturbed velocity gradient

$$\Delta l^{\varepsilon}_{(ij)} = \Delta F^{\varepsilon}_{(ij)} F^{-1} \tag{B.6}$$

and with (B.3) it follows

$$\Delta d^{\varepsilon}_{(ij)} = \frac{1}{2}(\Delta l_{(ij)} + \Delta l^{T}_{(ij)}) = \frac{\varepsilon}{2}(e_i \otimes e_j + e_j \otimes e_i) \tag{B.7}$$

$$\Delta w^{\varepsilon}_{(ij)} = \frac{1}{2}(\Delta l_{(ij)} - \Delta l^{T}_{(ij)}) = \mathbf{0} \quad . \tag{B.8}$$

Inserting (B.4) in (B.5) and using the relations of (B.7) and (B.8) one receives an expression in terms of the perturbed and unperturbed Kirchhoff stresses and the perturbation parameter only

$$\tau^{\varepsilon}_{(ij)} - \tau = \tilde{\mathbb{c}}^{\varepsilon}_{(ij)} : \frac{\varepsilon}{2}(e_i \otimes e_j + e_j \otimes e_i) \quad . \tag{B.9}$$

By make use of the symmetry properties the approximation of the spatial tangent moduli yields

$$\tilde{\mathbb{c}} \approx \tilde{\mathbb{c}}^{\varepsilon}_{(ij)} = \frac{1}{\varepsilon}(\tau^{\varepsilon}_{(ij)} - \tau) \tag{B.10}$$

and the approximated consistent spatial tangent regarding the Cauchy stress is given by

$$\tilde{\mathbb{C}}^{\varepsilon,\sigma}_{(ij)} = \frac{1}{J}\tilde{\mathbb{C}}^{\varepsilon}_{(ij)} \quad . \tag{B.11}$$

This equation results in a symmetric 6×6 matrix

$$\tilde{\mathbb{C}}^{\varepsilon,\sigma}_{(ij)} = \frac{1}{J\varepsilon}\begin{bmatrix} \tau_{11}(F^{\varepsilon}_{(11)}) - \tau_{11} & \dots & \tau_{11}(F^{\varepsilon}_{(23)}) - \tau_{11} \\ \tau_{22}(F^{\varepsilon}_{(11)}) - \tau_{22} & \dots & \tau_{22}(F^{\varepsilon}_{(23)}) - \tau_{22} \\ \tau_{33}(F^{\varepsilon}_{(11)}) - \tau_{33} & \dots & \tau_{33}(F^{\varepsilon}_{(23)}) - \tau_{33} \\ \tau_{12}(F^{\varepsilon}_{(11)}) - \tau_{12} & \dots & \tau_{12}(F^{\varepsilon}_{(23)}) - \tau_{12} \\ \tau_{13}(F^{\varepsilon}_{(11)}) - \tau_{13} & \dots & \tau_{13}(F^{\varepsilon}_{(23)}) - \tau_{13} \\ \tau_{23}(F^{\varepsilon}_{(11)}) - \tau_{23} & \dots & \tau_{23}(F^{\varepsilon}_{(23)}) - \tau_{23} \end{bmatrix} \tag{B.12}$$

where the components are the difference of the six perturbed Cauchy stresses resulting from the six perturbed deformation gradients and the unperturbed Cauchy stress with respect to the perturbation parameter ε. So, the numerical approximated tangent strongly depends on the choice of the perturbation parameter. Theoretically, taking this parameter very small the numerical tangent would converge to the analytical solution, but round-off errors of the computer avoid this convergence. Choosing the parameter too large results in an incorrect approximation. Rather, there must be an optimum of the parameter depending on the computer architecture and the problem.

As the analytical consistent tangent in Sec. 6.2.1 the numerically approximated tangent is computed with AceGen, too. The crucial advantage is that seven derivatives must be computed to get all stresses needed for the numerical tangent and that these derivatives are automatically computed by AceGen. As said before (cf. Ch. 6), the code will be optimized so that redundant components of the computed tensors are minimized. The algorithm to compute the tangent is presented in Box B.1.

Algorithm to compute the approximated spatial elasticity tensor with the FD method

input in UMAT: $\boldsymbol{F}_{t+\Delta t}$

→ computation with AceGen :

1. steps 1-4 of Box 6.1 to get Kirchhoff stress
2. perturbed deformation gradients: $\boldsymbol{F}^{\varepsilon}_{(ij)} = \boldsymbol{F} + \Delta \boldsymbol{F}^{\varepsilon}_{(ij)}$
3. perturbed right Cauchy-Green tensors: $\boldsymbol{C}^{\varepsilon}_{(ij)} = \boldsymbol{F}^{\varepsilon T}_{(ij)} \boldsymbol{F}^{\varepsilon}_{(ij)}$
4. strain energy functions with perturbed quantities: $W^{\varepsilon}_{(ij)}(\boldsymbol{C}^{\varepsilon}_{(ij)})$
5. perturbed Kirchhoff stress tensors (AD): $\boldsymbol{\tau}^{\varepsilon}_{(ij)} = 2\boldsymbol{F}^{\varepsilon}_{(ij)} \dfrac{\partial W(\boldsymbol{C}^{\varepsilon}_{(ij)})}{\partial \boldsymbol{C}^{\varepsilon}_{(ij)}} \boldsymbol{F}^{\varepsilon T}_{(ij)}$
6. approximated spatial elasticity tensor: $\tilde{\mathbb{C}}^{\varepsilon,\sigma}_{(ij)} \approx \dfrac{1}{J\varepsilon}(\boldsymbol{\tau}^{\varepsilon}_{(ij)} - \boldsymbol{\tau})$

Box B.1

B.1.2. Complex step derivative approximation

A further method to numerically approximate the tangent is the complex-step derivative approximation (CSDA) (Martins et al. (2003), Perez-Foguet et al. (2000)). The complex-step derivative approximation make use of the imaginary part of the perturbed state of a function. The perturbation parameter ε is multiplied by the imaginary number i

$$\frac{\mathrm{d}f(x)}{\mathrm{d}x} \approx \frac{\mathrm{Im}[f(x+\varepsilon i)]}{\varepsilon} \tag{B.13}$$

and the approximation does not suffer from round-off errors at small step size where the truncation error is small. Applying this approximation to the above description (e.g. Box B.1) only small changes must be performed in the algorithm workflow. However, AceGen is not able to handle complex numbers and is only used to compute the stress. The rest of the algorithm is performed using FORTRAN (Box B.2). So, the implementation of the CSDA is more complex and no closed coding with AceGen is possible.

Algorithm to compute the approximated spatial elasticity tensor with the CSDA method

input in UMAT: $\boldsymbol{F}_{t+\Delta t}$

→ computation with AceGen :

1. steps 1-4 of Box 6.1 to get Kirchhoff stress

→ computation with FORTRAN :

2. perturbed deformation gradients: $\boldsymbol{F}^{\varepsilon i}_{(ij)} = \boldsymbol{F} + \Delta \boldsymbol{F}^{\varepsilon}_{(ij)} i$
3. computation with subroutine of steps 1-4 of Box 6.1 to get perturbed Kirchhoff stress:
$$\boldsymbol{\tau}^{\varepsilon i}_{(ij)} = 2\boldsymbol{F}^{\varepsilon i}_{(ij)} \frac{\partial W(\boldsymbol{C}^{\varepsilon i}_{(ij)})}{\partial \boldsymbol{C}^{\varepsilon i}_{(ij)}} \boldsymbol{F}^{\varepsilon i T}_{(ij)}$$
4. approximated spatial elasticity tensor: $\tilde{\mathbb{c}}^{\varepsilon,\sigma}_{(ij)} \approx \dfrac{\mathrm{Im}[\boldsymbol{\tau}^{\varepsilon i}_{(ij)}]}{J\varepsilon}$

Box B.2

B.1.3. Additional discussion of computations of Sec. 6.2.1

In this section an additional discussion to the verification section of Sec. 6.2.1 regarding the convergence behavior of the tangent derived by the numerical approximation methods is given. The same neo-Hookean strain energy function is used and the same computational models are used. Again, the number of iterations and the convergence rate are compared with the analytical solution of the tangent.Four different routines are generated with AceGen: the analytical, AD, FD and CSDA tangent. In Tab. B.1 these routines are listed with the corresponding size of each routine in bytes. For the approximation of the tangent using the CSDA method only the stress is computed with AceGen. This gives a small subroutine but it must be called seven times by the main routine to compute the unperturbed and perturbed stresses.

Single element test

The relative errors η (6.12) of the computed tangents with respect to the analytic (reference) solution ${}^{ref}\tilde{\mathbb{c}}^{\sigma}$ are evaluated in a single element test as done before in Sec. 6.2.1. The errors are evaluated in the first iteration and are shown in Fig. B.1. With the automatic

Table B.1.: By AceGen generated routines for a UMAT

analytical tangent	reference solution	2938 bytes
tangent by AceGen	AD (Box 6.1)	25813 bytes
numerical tangent by AceGen	FD (Box B.1)	16421 bytes
numerical tangent by AceGen (only stress)	CSDA (Box B.2)	1959 bytes

differentiation method (AD) a very small error occurs ($\eta \approx 10^{-16}$) which is in the region of the machine precision of a 64bit architecture ($2.2 \cdot 10^{-16}$). The error of the numerical approximation using the forward difference approximation (FD) decreases linearly with decreasing step size due to decreasing truncation error. At a step size of $\varepsilon = 10^{-8}$ a minimal error occurs and with further decreasing step size the error increases due to round-off errors. The error of the complex-step derivative approximation (CSDA) of the tangent converges quadratically to a error near to that of the AD method at a step size of $\varepsilon = 10^{-8}$. No error is found at a step size of $\varepsilon = 10^{-15} - 10^{-16}$ where the approximation is equal to the reference solution. The reference (analytical) solution needs six iterations to converge.

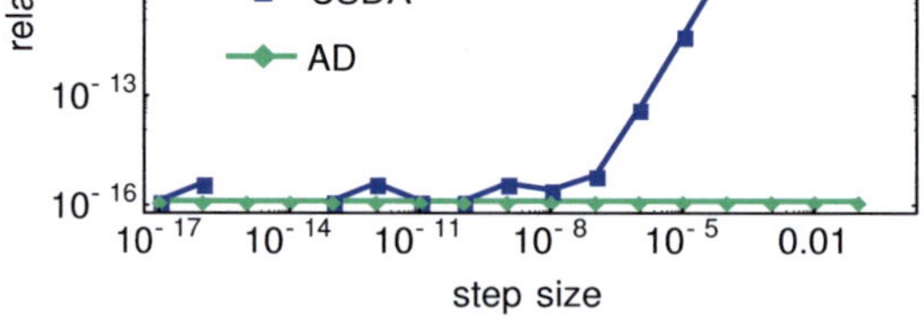

Figure B.1.: Relative errors of computed tangents in the single element test

The convergence behavior of the three other methods in terms of the norm of the largest residual force is shown in Tab. B.2. The tangent derived by the AD method gives the same convergence as the analytical one which is not surprising because the relative error is very small. In case of FD approximation the largest residual norm is the same to the analytical solution in the step size range of $\varepsilon = 10^{-6} - 10^{-12}$ and in case of the CSDA good results are achieved even from a step size of $\varepsilon = 10^{-4}$. Remarkable is, that the CSDA gives the same residual force in the last iteration in contrast to the FD approximation. The computation of the stress at the end of the simulation gives same values for every method.

Simulation of three point bending test

Based on the results of the single element test, a step size of $\varepsilon_{FD} = 10^{-8}$ for the FD method and $\varepsilon_{CSDA} = 10^{-16}$ for the CSDA method is chosen for the simulation of a three point bending experiment. The number of iterations are in every simulation the same as well as a quadratic convergence rate is obtained near the solution in all cases (Tab. B.3).

This behavior shows that the computation of the tangent is in all three cases right or well enough approximated. Again, evaluating the relative error of the respective three tangents in the third increment of an element under a complex stress state, nearly the same behavior as in the single element test is observed (Fig. B.2).

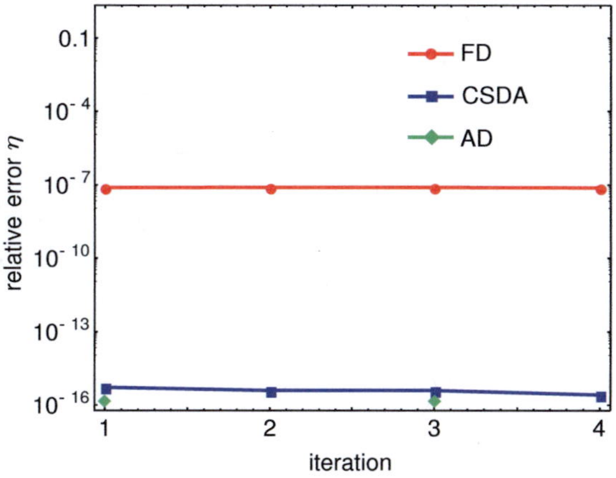

Figure B.2.: Relative error of computed tangents in the third increment of the three point bending simulation

Table B.2.: Norm of largest residual force

iteration	analytical tangent	AD tangent	FD $\varepsilon = 10^{-2}$	CSDA $\varepsilon = 10^{-2}$	FD $\varepsilon = 10^{-4}$	CSDA $\varepsilon = 10^{-4}$
1	$4.447 \cdot 10^4$	$4.447 \cdot 10^4$	$8.355 \cdot 10^4$	$4.485 \cdot 10^4$	$4.474 \cdot 10^4$	$4.447 \cdot 10^4$
2	$1.961 \cdot 10^4$	$1.961 \cdot 10^4$	$3.592 \cdot 10^4$	$1.972 \cdot 10^4$	$1.972 \cdot 10^4$	$1.961 \cdot 10^4$
3	$7.741 \cdot 10^3$	$7.741 \cdot 10^3$	$1.520 \cdot 10^4$	$7.772 \cdot 10^3$	$7.797 \cdot 10^3$	$7.741 \cdot 10^3$
4	815	815	$4.386 \cdot 10^3$	807	837	815
5	7.27	7.27	367	8.26	7.96	7.27
6	$6.554 \cdot 10^{-4}$	$6.554 \cdot 10^{-4}$	136	$9.248 \cdot 10^{-3}$	$5.586 \cdot 10^{-3}$	$6.545 \cdot 10^{-4}$
7			4.50			
8			0.256			

iteration	FD $\varepsilon = 10^{-6}$	CSDA $\varepsilon = 10^{-6}$	FD $\varepsilon = 10^{-8}$	CSDA $\varepsilon = 10^{-8}$	FD $\varepsilon = 10^{-10}$	CSDA $\varepsilon = 10^{-10}$
1	$4.447 \cdot 10^4$	$4.447 \cdot 10^4$	$4.447 \cdot 10^4$	$4.447 \cdot 10^4$	$4.447 \cdot 10^4$	$4.447 \cdot 10^4$
2	$1.961 \cdot 10^4$	$1.961 \cdot 10^4$	$1.961 \cdot 10^4$	$1.961 \cdot 10^4$	$1.961 \cdot 10^4$	$1.961 \cdot 10^4$
3	$7.741 \cdot 10^3$	$7.741 \cdot 10^3$	$7.741 \cdot 10^3$	$7.741 \cdot 10^3$	$7.741 \cdot 10^3$	$7.741 \cdot 10^3$
4	815	815	815	815	815	815
5	7.27	7.27	7.27	7.27	7.27	7.27
6	$7.173 \cdot 10^{-4}$	$6.554 \cdot 10^{-4}$	$6.560 \cdot 10^{-4}$	$6.554 \cdot 10^{-4}$	$6.564 \cdot 10^{-4}$	$6.554 \cdot 10^{-4}$

iteration	FD $\varepsilon = 10^{-12}$	CSDA $\varepsilon = 10^{-12}$	FD $\varepsilon = 10^{-14}$	CSDA $\varepsilon = 10^{-14}$	FD $\varepsilon = 10^{-16}$	CSDA $\varepsilon = 10^{-16}$
1	$4.447 \cdot 10^4$	$4.447 \cdot 10^4$	$4.451 \cdot 10^4$	$4.447 \cdot 10^4$	$3.509 \cdot 10^4$	$4.447 \cdot 10^4$
2	$1.961 \cdot 10^4$	$1.961 \cdot 10^4$	$1.964 \cdot 10^4$	$1.961 \cdot 10^4$	$1.616 \cdot 10^4$	$1.961 \cdot 10^4$
3	$7.741 \cdot 10^3$	$7.741 \cdot 10^3$	$7.751 \cdot 10^3$	$7.741 \cdot 10^3$	$6.000 \cdot 10^3$	$7.741 \cdot 10^3$
4	815	815	822	815	837	815
5	7.27	7.27	9.04	7.27	144	7.27
6	$7.016 \cdot 10^{-4}$	$6.554 \cdot 10^{-4}$	$1.139 \cdot 10^{-2}$	$6.554 \cdot 10^{-4}$	16.7	$6.554 \cdot 10^{-4}$
7					1.49	

Table B.3.: Convergence behavior of the simulation. The norm of the largest residual force is given in the third increment.

	analytical tangent	AD tangent	FD ($\varepsilon = 10^{-8}$)	CSDA ($\varepsilon = 10^{-16}$)
total no. of iterations	22	22	22	22
normalized computation time	1.0	1.0278	1.0085	1.0961
iteration #				
1	14.9	14.9	14.9	14.9
2	5.41	5.41	5.41	5.41
3	$1.705 \cdot 10^{-2}$	$1.705 \cdot 10^{-2}$	$1.705 \cdot 10^{-2}$	$1.705 \cdot 10^{-2}$
4	$8.687 \cdot 10^{-6}$	$8.687 \cdot 10^{-6}$	$8.687 \cdot 10^{-6}$	$4.835 \cdot 10^{-8}$

The computation time (Tab. 6.3) differs from routine to routine and correlates with the size of the routines (Tab. 6.1). Of course, the fastest routine is the analytical one because the fewest operations must be performed due to the direct implementation of the derivatives. The computation time of the numerical approximation with the FD method are very fast as well. More operations must be performed with the tangent computed by AceGen and this leads to a longer computation time. The longest computation time is needed by the routine with the CSDA method because the only optimized code is that of the computation of the stress. The rest of the code is programmed by hand because of the inability of AceGen to work with complex numbers.

B.2. Numerical approximation of algorithmic consistent tangent moduli

The approximation of the algorithmic consistent tangent is similar to the methods of Sec. B.1 but in case of inelasticity the update algorithm of the internal variables must be considered (Miehe, 1996). Introducing a set of internal variables $\boldsymbol{H}$ which are updated through an update algorithm $\boldsymbol{H}^{alg}$ which gives the updated internal variables $\boldsymbol{H}_{t+\Delta t} =$

$\boldsymbol{H}^{alg}(\boldsymbol{F}_{t+\Delta t},\boldsymbol{H}_t)$ at the end of the time step. In this case the stress depends on these updated variables and the total deformation as

$$\boldsymbol{\tau}_{t+\Delta t} = \boldsymbol{\tau}(\boldsymbol{F}_{t+\Delta t},\boldsymbol{H}_{t+\Delta t}) = \boldsymbol{\tau}^{alg}(\boldsymbol{F}_{t+\Delta t},\boldsymbol{H}_t) \quad . \tag{B.14}$$

Using the same perturbation strategy as in Sec. B.1 the perturbed stress with respect to the update algorithm yields

$$\boldsymbol{\tau}^{\varepsilon}_{(i,j)} = \boldsymbol{\tau}^{\varepsilon,alg}_{(i,j)}(\boldsymbol{F}^{\varepsilon}_{(i,j)},\boldsymbol{H}_t) \tag{B.15}$$

by suppressing the time step subscripts for convenience. Applying the forward difference approximation (FD) (cf. Sec. B.1) the approximated consistent tangent yields

$$\tilde{\mathbb{C}}^{alg,\sigma} \approx \tilde{\mathbb{C}}^{\varepsilon,alg,\sigma}_{(ij)} = \frac{1}{J\varepsilon}(\boldsymbol{\tau}^{\varepsilon,alg}_{(i,j)}(\boldsymbol{F}^{\varepsilon}_{(i,j)},\boldsymbol{H}_t) - \boldsymbol{\tau}^{alg}(\boldsymbol{F},\boldsymbol{H}_t)) \tag{B.16}$$

for all quantities at time $t + \Delta t$ otherwise noted. Applying the complex step derivative approximation (CSDA) the approximation of the consistent tangent is given by

$$\tilde{\mathbb{C}}^{alg,\sigma} \approx \tilde{\mathbb{C}}^{\varepsilon,alg,\sigma}_{(ij)} = \frac{1}{J\varepsilon}\mathrm{Im}[\boldsymbol{\tau}^{\varepsilon i,alg}_{(i,j)}(\boldsymbol{F}^{\varepsilon i}_{(i,j)},\boldsymbol{H}_t)] \quad . \tag{B.17}$$

The algorithms of both methods are presented with respect to a elastic-plastic model in Box B.3 and Box B.4, respectively.

Algorithm to compute the approximated consistent tangent with the FD method

input in UMAT: $\boldsymbol{F}_{t+\Delta t},\boldsymbol{H}_t$

→ computation with AceGen :

1. perform algorithm of Box 6.3 (till step 2.2) with $\boldsymbol{H} = [\boldsymbol{F}^p \ \varepsilon^p]$

   to get updated quantities $\boldsymbol{\tau}_{t+\Delta t},\boldsymbol{H}_{t+\Delta t}$
2. perturbation of deformation gradients: $\boldsymbol{F}^{\varepsilon}_{(ij)} = \boldsymbol{F} + \Delta\boldsymbol{F}^{\varepsilon}_{(ij)}$
3. perform for each perturbed deformation gradient the algorithm of Box 6.3 (till step 2.2)

   to get perturbed algorithmic Kirchhoff stress $\boldsymbol{\tau}^{\varepsilon,alg}_{(i,j)}(\boldsymbol{F}^{\varepsilon}_{(i,j)},\boldsymbol{H}_t)$
4. approximated consistent tangent moduli: $\tilde{\mathbb{C}}^{\varepsilon,alg,\sigma}_{(ij)} \approx \frac{1}{J\varepsilon}(\boldsymbol{\tau}^{\varepsilon,alg}_{(ij)}(\boldsymbol{F}^{\varepsilon}_{(i,j)},\boldsymbol{H}_t) - \boldsymbol{\tau}_{t+\Delta t})$

Box B.3

Algorithm to compute the approximated consistent tangent with the CSDA method

input in UMAT: $\boldsymbol{F}_{t+\Delta t}, \boldsymbol{H}_t$

$\rightarrow$ computation with AceGen :

1. perform algorithm of Box 6.3 (till step 2.2) with $\boldsymbol{H} = [\boldsymbol{F}^p \; \varepsilon^p]$

   to get updated quantities $\boldsymbol{\tau}_{t+\Delta t}, \boldsymbol{H}_{t+\Delta t}$

$\rightarrow$ computation with FORTRAN :

2. perturbation of deformation gradients: $\boldsymbol{F}^{\varepsilon i}_{(ij)} = \boldsymbol{F} + \Delta \boldsymbol{F}^{\varepsilon}_{(ij)} i$
3. perform for each perturbed deformation gradient the algorithm of Box 6.3 (till step 2.2)

   to get perturbed algorithmic Kirchhoff stress $\boldsymbol{\tau}^{\varepsilon,alg}_{(i,j)}(\boldsymbol{F}^{\varepsilon}_{(i,j)}, \boldsymbol{H}_t)$
4. approximated consistent tangent moduli: $\tilde{\mathbb{c}}^{\varepsilon,alg,\sigma}_{(ij)} = \dfrac{\mathrm{Im}[\boldsymbol{\tau}^{\varepsilon i,alg}_{(ij)}(\boldsymbol{F}^{\varepsilon i}_{(i,j)}, \boldsymbol{H}_t)]}{J\varepsilon}$

Box B.4

B.2.1. Additional discussion of computations of Sec. 6.2.2

As the hyperelastic material model implementation is additionally discussed in Sec. B.1.3 with respect to the convergence behavior of the approximated tangents, the elastic-plastic model implementation is discussed in this section as well. For the used material model and the computational models it is referred to Sec. 6.2.2. The generated subroutines are listed with their size in bytes in Tab. B.4. Again, for the CSDA method only the routine for the update of the internal variables and stress is used. This leads to seven executions of the subroutine.

Table B.4.: With AceGen generated routines for UMAT

tangent by AceGen	AD (Box 6.3)	309589 bytes
numerical tangent by AceGen	FD (Box B.3)	1080376 bytes
numerical tangent by AceGen (only stress)	CSDA (Box B.4)	140699 bytes

All three methods to compute the tangent result in the same convergence behavior (Tab. B.5) in the simulation of a cylindrical bar. All need the same number of iterations but differ in the computation time dramatically. While the routine generated by AD is the fastest one, the CSDA method is the slowest one with a factor of almost two with respect to the AD. This result is not surprising since AceGen in combination with the AD method generates smallest source code (cf. Tab. B.4) while for the CSDA method only the stress update routine is generated by AceGen which must be called seven times. This decelerates the computation time of course. In contrast, the FD method is completely coded with AceGen so that the optimization of the code is fully exploited which results in a slower computation time of a factor of 1.3 with respect to the routine coded with AD.

Table B.5.: Convergence behavior of the simulation. The norm of the largest residual force is given in the last increment.

	AD (AceGen)	FD ($\varepsilon = 10^{-8}$)	CSDA ($\varepsilon = 10^{-16}$)
total no. of iterations	227	227	227
normed computation wall-clock time	1.0	1.345	1.958
iteration #			
1	0.430	0.430	0.430
2	0.128	0.128	0.128
3	$3.606 \cdot 10^{-2}$	$3.606 \cdot 10^{-2}$	$3.606 \cdot 10^{-2}$
4	$9.177 \cdot 10^{-3}$	$9.177 \cdot 10^{-3}$	$9.177 \cdot 10^{-3}$
5	$1.570 \cdot 10^{-3}$	$1.569 \cdot 10^{-3}$	$1.569 \cdot 10^{-3}$

C. Complete AceGen codes

C.1. Hyperelastic material model implementation

AceGen and Mathematica code

```
Clear["Global`*"]
Get["AceGen`"]

SMSInitialize["routine", "Language" → "Fortran", "Mode" → "Optimal"];
SMSModule["routine", Real[stress$$[3, 3], tangent$$[6, 6],
    defgrad$$[3, 3], C10$$, D1$$]];

𝔽 ⊨ Array[SMSReal[defgrad$$[#1, #2]] &, {3, 3}];
C10 ⊨ SMSReal[C10$$];
D1 ⊨ SMSReal[D1$$];

tovoigt[tensor_] :=
  ({{tensor⟦1, 1, 1, 1⟧, tensor⟦1, 1, 2, 2⟧, tensor⟦1, 1, 3, 3⟧,
     tensor⟦1, 1, 1, 2⟧, tensor⟦1, 1, 1, 3⟧, tensor⟦1, 1, 2, 3⟧},
    {tensor⟦2, 2, 1, 1⟧, tensor⟦2, 2, 2, 2⟧, tensor⟦2, 2, 3, 3⟧,
     tensor⟦2, 2, 1, 2⟧, tensor⟦2, 2, 1, 3⟧, tensor⟦2, 2, 2, 3⟧},
    {tensor⟦3, 3, 1, 1⟧, tensor⟦3, 3, 2, 2⟧, tensor⟦3, 3, 3, 3⟧,
     tensor⟦3, 3, 1, 2⟧, tensor⟦3, 3, 1, 3⟧, tensor⟦3, 3, 2, 3⟧},
    {tensor⟦1, 2, 1, 1⟧, tensor⟦1, 2, 2, 2⟧, tensor⟦1, 2, 3, 3⟧,
     tensor⟦1, 2, 1, 2⟧, tensor⟦1, 2, 1, 3⟧, tensor⟦1, 2, 2, 3⟧},
    {tensor⟦1, 3, 1, 1⟧, tensor⟦1, 3, 2, 2⟧, tensor⟦1, 3, 3, 3⟧,
     tensor⟦1, 3, 1, 2⟧, tensor⟦1, 3, 1, 3⟧, tensor⟦1, 3, 2, 3⟧},
    {tensor⟦2, 3, 1, 1⟧, tensor⟦2, 3, 2, 2⟧, tensor⟦2, 3, 3, 3⟧,
     tensor⟦2, 3, 1, 2⟧, tensor⟦2, 3, 1, 3⟧, tensor⟦2, 3, 2, 3⟧}});
```

```
SMSFreeze[ℂ, Transpose[𝔽].𝔽, "Symmetric" → True];
𝕀ℂ ⊨ SMSInvariantsI[ℂ];
W ⊨ C10 * (𝕀ℂ⟦3⟧^(-1/3) * 𝕀ℂ⟦1⟧ - 3) + 1/D1 * (SMSSqrt[𝕀ℂ⟦3⟧] - 1)^2;
dWdℂ ⊨ SMSD[W, ℂ, "Symmetric" → True];
τ ⊨ 2 * 𝔽.dWdℂ.Transpose[𝔽];
J ⊨ SMSDet[𝔽];
σ ⊨ τ / J;
C ⊨ 4 * SMSD[dWdℂ, ℂ, "Symmetric" → True];

𝕔 ⫤ Table[0, {i, 3}, {j, 3}, {k, 3}, {l, 3}];
𝕒 ⫤ Table[0, {i, 3}, {j, 3}, {k, 3}, {l, 3}];

Do[Do[Do[Do[Do[Do[Do[Do[𝕔⟦i, j, k, l⟧ = 𝕔⟦i, j, k, l⟧
            + 𝔽⟦i, ii⟧ * 𝔽⟦j, jj⟧ * 𝔽⟦k, kk⟧ * 𝔽⟦l, ll⟧
             * C⟦ii, jj, kk, ll⟧, {ll, 3}], {kk, 3}], {jj, 3}],
      {ii, 3}], {l, 3}], {k, 3}], {j, 3}], {i, 3}];

Do[Do[Do[Do[𝕒⟦i, j, k, l⟧ = 1/2 * (SMSKroneckerDelta[i, k]
          * τ⟦j, l⟧ + τ⟦i, k⟧ * SMSKroneckerDelta[j, l]
         + SMSKroneckerDelta[i, l] * τ⟦j, k⟧ + τ⟦i, l⟧
          * SMSKroneckerDelta[j, k]), {l, 3}], {k, 3}],
   {j, 3}], {i, 3}];

𝕔alg ⊨ (tovoigt[𝕔] + tovoigt[𝕒]) / J;

SMSExport[σ, stress$$];
SMSExport[𝕔alg, tangent$$];
SMSWrite["IncludeHeaders" → {}, "LocalAuxiliaryVariables" → True];
```

C.2. Elastic-plastic material model implementation

AceGen and Mathematica code

```
Clear["Global`*"]
Get["AceGen`"]
SMSInitialize["plastic", "Language" → "Fortran", "Mode" → "Optimal",
  "VectorLength" → 100 000];
SMSModule["plastic", Real[F1$$[3, 3], Fp$$[3, 3], stress$$[3, 3],
   tangent$$[6, 6], tau0$$, k$$, E$$, mu$$, epsp$$]];
𝔽pt0 ⊧ Array[SMSReal[Fp$$[#1, #2]] &, {3, 3}];
𝔽t1 ⊧ Array[SMSReal[F1$$[#1, #2]] &, {3, 3}];
{τ0, k, Emodul, ν, ϵpt0} ⊧ SMSReal[{tau0$$, k$$, E$$, mu$$, epsp$$}];
Δλt0 ⊧ 0;
{λ, μ} ⊧ SMSHookeToLame[Emodul, ν];

𝔽etr ⊧ 𝔽t1.SMSInverse[𝔽pt0];
SMSFreeze[ℂe, Transpose[𝔽etr].𝔽etr,
  "Ignore" → (# === 0 &), "Symmetric" → True];
𝕀ℂe ⊧ SMSInvariantsI[ℂe];
We ⊧ μ / 2 * (𝕀ℂe[[1]] - 3) - μ / 2 * Log[𝕀ℂe[[3]]] + λ / 2 * (SMSSqrt[𝕀ℂe[[3]]] - 1) ^ 2;
τtrial ⊧ 2 * 𝔽etr.SMSD[We, ℂe, "IgnoreNumbers" → True,
     "Symmetric" → True].Transpose[𝔽etr];
s ⊧ τtrial - 1 / 3 Tr[τtrial] * IdentityMatrix[3];
τM ⊧ SMSSqrt[Tr[s.s]];
Wp ⊧ 1 / 2 * k * ϵpt0 ^ 2;
q ⊧ SMSD[Wp, ϵpt0];
f ⊧ τM - SMSSqrt[2 / 3] * (τ0 + q);

Ht0 ⊧ ({𝔽pt0[[1, 1]], 𝔽pt0[[1, 2]], 𝔽pt0[[1, 3]], 𝔽pt0[[2, 1]], 𝔽pt0[[2, 2]],
    𝔽pt0[[2, 3]], 𝔽pt0[[3, 1]], 𝔽pt0[[3, 2]], 𝔽pt0[[3, 3]], Δλt0, ϵpt0});

SMSIf[SMSLogical[f < 10^-8]];
Ht1 ⊣ Ht0;
```

```
SMSElse[];
Xi ⊣ Ht0;
SMSDo[ii, 1, 30, 1, Xi];
 𝔽pi ⊨ ({{Xi⟦1⟧, Xi⟦2⟧, Xi⟦3⟧}, {Xi⟦4⟧, Xi⟦5⟧, Xi⟦6⟧},
     {Xi⟦7⟧, Xi⟦8⟧, Xi⟦9⟧}});
 Δλi ⊨ Xi⟦10⟧;
 SMSFreeze[εpi, Xi⟦11⟧];
 𝔽ei ⊨ 𝔽t1.SMSInverse[𝔽pi];
 SMSFreeze[ℂei, Transpose[𝔽ei].𝔽ei, "Symmetric" → True];
 𝕀ℂei ⊨ SMSInvariantsI[ℂei];
 Wei ⊨ μ / 2 * (𝕀ℂei⟦1⟧ - 3) - μ / 2 * Log[𝕀ℂei⟦3⟧]
+ λ / 2 * (SMSSqrt[𝕀ℂei⟦3⟧] - 1) ^ 2;
 SMSFreeze[τi, 2 * 𝔽ei.SMSD[Wei, ℂei, "IgnoreNumbers" → True,
      "Symmetric" → True].Transpose[𝔽ei], "KeepStructure" → True];
 𝕤i ⊨ τi - 1 / 3 Tr[τi] IdentityMatrix[3];
 τMi ⊨ SMSSqrt[Tr[𝕤i.𝕤i]];
 Wpi ⊨ 1 / 2 * k * εpi ^ 2;
 SMSFreeze[qi, SMSD[Wpi, εpi]];
 fi ⊨ τMi - SMSSqrt[2 / 3] * (τ0 + qi);
 ni ⊨ Simplify[SMSD[fi, τi, "IgnoreNumbers" → True,
     "Symmetric" → True]];
 hi ⊨ -SMSD[fi, qi];
 R1i ⊨ Simplify[𝔽ei - SMSMatrixExp[-Δλi * ni, "Order" → 10,
       "Module" → False].𝔽etr];
 R2i ⊨ εpi - εpt0 - Δλi * hi;
 R3i ⊨ fi;
 Ri ⊨ {R1i⟦1, 1⟧, R1i⟦1, 2⟧, R1i⟦1, 3⟧, R1i⟦2, 1⟧, R1i⟦2, 2⟧,
    R1i⟦2, 3⟧, R1i⟦3, 1⟧, R1i⟦3, 2⟧, R1i⟦3, 3⟧, R2i, R3i};
 RX ⊨ SMSD[Ri, Xi];
 LU ⊣ SMSLUFactor[RX];
 ΔX ⊨ SMSLUSolve[LU, -Ri];
 Xi ⊣ Xi + ΔX;
 ΔXN ⊨ SMSSqrt[ΔX.ΔX / Length[ΔX]];
 SMSIf[ΔXN < 10 ^ -8];
  dRd𝔽 ⊨ SMSD[Ri, Flatten[𝔽t1], "Constant" → Xi];
  dXd𝔽 ⊣ SMSLUSolve[LU, -dRd𝔽];
  SMSBreak[];
 SMSEndIf[];
SMSEndDo[Xi, dXd𝔽];
Ht1 ⊣ SMSFreeze[Xi, "Dependency" → {Flatten[𝔽t1], dXd𝔽}];
SMSEndIf[Ht1];
𝔽pt1 ⊨ ({{Ht1⟦1⟧, Ht1⟦2⟧, Ht1⟦3⟧}, {Ht1⟦4⟧, Ht1⟦5⟧, Ht1⟦6⟧},
     {Ht1⟦7⟧, Ht1⟦8⟧, Ht1⟦9⟧}});
εpt1 ⊨ Ht1⟦11⟧;
```

```
𝔽et1 ⊧ 𝔽t1.SMSInverse[𝔽pt1];
J ⊧ SMSDet[𝔽t1];
SMSFreeze[ℂe1, Transpose[𝔽et1].𝔽et1, "Symmetric" → True,
  "KeepStructure" → True];
𝕀ℂe1 ⊧ SMSInvariantsI[ℂe1];
We1 ⊧ μ / 2 * (𝕀ℂe1⟦1⟧ - 3) - μ / 2 * Log[𝕀ℂe1⟦3⟧]
+ λ / 2 * (SMSSqrt[𝕀ℂe1⟦3⟧] - 1) ^ 2;
S1 ⊧ 2 * SMSD[We1, ℂe1, "IgnoreNumbers" → True,
    "Symmetric" → True, "Constant" → Ht1];
τ1 ⊧ 𝔽et1.S1.Transpose[𝔽et1];
σ1 ⊧ τ1 / J;

tovoigt4[tensor_] :=
  ({{tensor⟦1, 1, 1, 1⟧, tensor⟦1, 1, 2, 2⟧, tensor⟦1, 1, 3, 3⟧,
     tensor⟦1, 1, 1, 2⟧, tensor⟦1, 1, 1, 3⟧, tensor⟦1, 1, 2, 3⟧},
    {tensor⟦2, 2, 1, 1⟧, tensor⟦2, 2, 2, 2⟧, tensor⟦2, 2, 3, 3⟧,
     tensor⟦2, 2, 1, 2⟧, tensor⟦2, 2, 1, 3⟧, tensor⟦2, 2, 2, 3⟧},
    {tensor⟦3, 3, 1, 1⟧, tensor⟦3, 3, 2, 2⟧, tensor⟦3, 3, 3, 3⟧,
     tensor⟦3, 3, 1, 2⟧, tensor⟦3, 3, 1, 3⟧, tensor⟦3, 3, 2, 3⟧},
    {tensor⟦1, 2, 1, 1⟧, tensor⟦1, 2, 2, 2⟧, tensor⟦1, 2, 3, 3⟧,
     tensor⟦1, 2, 1, 2⟧, tensor⟦1, 2, 1, 3⟧, tensor⟦1, 2, 2, 3⟧},
    {tensor⟦1, 3, 1, 1⟧, tensor⟦1, 3, 2, 2⟧, tensor⟦1, 3, 3, 3⟧,
     tensor⟦1, 3, 1, 2⟧, tensor⟦1, 3, 1, 3⟧, tensor⟦1, 3, 2, 3⟧},
    {tensor⟦2, 3, 1, 1⟧, tensor⟦2, 3, 2, 2⟧, tensor⟦2, 3, 3, 3⟧,
     tensor⟦2, 3, 1, 2⟧, tensor⟦2, 3, 1, 3⟧, tensor⟦2, 3, 2, 3⟧}});
```

```
SMSIf[SMSLogical[f < 10^-8]];

ℭel ⊨ 2 SMSD[S1, ℭel, "IgnoreNumbers" → True, "Symmetric" → True];
𝕔el ⫤ Table[0, {i, 3}, {j, 3}, {k, 3}, {l, 3}];
Do[Do[Do[Do[Do[Do[Do[Do[𝕔el⟦i, j, k, l⟧ = 𝕔el⟦i, j, k, l⟧ +
            𝔽et1⟦i, ii⟧ * 𝔽et1⟦j, jj⟧ * 𝔽et1⟦k, kk⟧ * 𝔽et1⟦l, ll⟧
             * ℭel⟦ii, jj, kk, ll⟧, {ll, 3}], {kk, 3}], {jj, 3}],
      {ii, 3}], {l, 3}], {k, 3}], {j, 3}], {i, 3}];

𝕒el ⫤ Table[0, {i, 3}, {j, 3}, {k, 3}, {l, 3}];
Do[Do[Do[Do[𝕒el⟦i, j, k, l⟧ = 𝕒el⟦i, j, k, l⟧
        + 1 / 2 * (SMSKroneckerDelta[i, k] * τ1⟦j, l⟧ + τ1⟦i, k⟧
            * SMSKroneckerDelta[j, l] + SMSKroneckerDelta[i, l]
            * τ1⟦j, k⟧ + τ1⟦i, l⟧ * SMSKroneckerDelta[j, k]), {l, 3}],
     {k, 3}], {j, 3}], {i, 3}];
𝕔alg ⫤ tovoigt4[𝕔el + 𝕒el] / J;

SMSElse[];

𝕔ep ⊨ SMSD[τ1, 𝔽t1, "IgnoreNumbers" → True].Transpose[𝔽t1];
𝕒ep ⫤ Table[0, {i, 3}, {j, 3}, {k, 3}, {l, 3}];
Do[Do[Do[Do[𝕒ep⟦i, j, k, l⟧ = 𝕒ep⟦i, j, k, l⟧
        + 1 / 2 (-SMSKroneckerDelta[i, k] * τ1⟦j, l⟧ + τ1⟦i, k⟧
            * SMSKroneckerDelta[j, l] + SMSKroneckerDelta[i, l]
            * τ1⟦j, k⟧ - τ1⟦i, l⟧ * SMSKroneckerDelta[j, k]), {l, 3}],
     {k, 3}], {j, 3}], {i, 3}];

𝕔alg ⊣ tovoigt4[𝕔ep + 𝕒ep] / J;
SMSEndIf[𝕔alg];

SMSExport[σ1, stress$$];
SMSExport[𝕔alg, tangent$$];
SMSExport[𝔽pt1, Fp$$];
SMSExport[ϵpt1, epsp$$];
SMSWrite["IncludeHeaders" → {}, "LocalAuxiliaryVariables" → True];
```

Schriftenreihe des Instituts für Mechanik

ISSN 2363-4936

Die Bände sind unter www.ksp.kit.edu als PDF frei verfügbar oder als Druckausgabe bestellbar.

Band 1 Marlon Franke
Discretisation techniques for large deformation computational contact elastodynamics. 2014
ISBN 978-3-7315-0278-4

Band 2 Philipp Hempel
Constitutive modeling of amorphous thermoplastic polymers with special emphasis on manufacturing processes. 2016
ISBN 978-3-7315-0550-1